Woodward

Understanding Polymer Morphology

Hanser **Understanding** Books
A Series of Mini-Tutorials

Series Editor: E.H. Immergut

Understanding Injection Molding Technology (Rees)
Understanding Polymer Morphology (Woodward)

In Preparation:

Understanding Mold Design (Gerson)
Understanding Quality Systems for Polymer Processing (Keating)
Understanding Adhesion and Adhesives Technology (Pocius)

Arthur E. Woodward

Understanding
Polymer Morphology

Hanser Publishers, Munich Vienna New York

Hanser/Gardner Publications, Inc., Cincinnati

The Author:
Arthur E. Woodward, 70 E. 10th Street, New York, NY 10003, USA

Distributed in the USA and in Canada by
Hanser/Gardner Publications, Inc.
6600 Clough Pike, Cincinnati, Ohio 45244-4090, USA
Fax: (513) 527-8950
Phone: (513) 527-8977 or (800) 950-8977

Distributed in all other countries by
Carl Hanser Verlag
Postfach 86 04 20, 81631 München, Germany
Fax: +49 (89) 98 48 09

The use of general descriptive names, trademarks, etc., in this publication, even if the former are not especially identified, is not to be taken as a sign that such names, as understood by the Trade Marks and Merchandise Marks Act, may accordingly be used freely by anyone.

While the advice and information in this book are believed to be true and accurate at the date of going to press, neither the authors nor the editors nor the publisher can accept any legal responsibility for any errors or omissions that may be made. The publisher makes no warranty, express or implied, with respect to the material contained herein.

Library of Congress Cataloging-in-Publication Data
Woodward, Arthur E., 1925-
    Understanding polymer morphology / Arthur E. Woodward,
        p.   cm. – (Hanser understanding books)
    Includes bibliographical references and index.
    ISBN 1-56990-141-4
    1. Polymers.   I. Title.   II. Series.
TP 1087.W66   1995
668.9--dc20   94-37332

Die Deutsche Bibliothek-CIP-Einheitsaufnahme
Woodward, Arthur E.:
Understanding polymer morphology / Arthur E. Woodward. -
Munich ; Vienna ; New York : Hanser ; Cincinnati : Hanser/
Gardner, 1995
    (Understanding)
ISBN 3-446-17431-1

© Carl Hanser Verlag, Munich Vienna New York, 1995
Typeset in the USA by pageAbility, Watertown
Printed and bound in Germany by Schoder Druck GmbH & Co. KG, Gersthofen

Introduction to the Series

In order to keep up in today's world of rapidly changing technology we need to open our eyes and ears and, most importantly, our minds to new scientific ideas and methods, new engineering approaches and manufacturing technologies and new product design and applications. As students graduate from college and either pursue academic polymer research or start their careers in the plastics industry, they are exposed to problems, materials, instruments and machines that are unfamiliar to them. Similarly, many working scientists and engineers who change jobs must quickly get up to speed in their new environment.

To satisfy the needs of these "newcomers" to various fields of polymer science and plastics engineering, we have invited a number of scientists and engineers, who are experts in their field and also good communicators, to write short, introductory books which let the reader "understand" the topic rather than to overwhelm him/her with a mass of facts and data. We have encouraged our authors to write the kind of book that can be read profitably by a beginner, such as a new company employee or a student, but also by someone familiar with the subject, who will gain new insights and a new perspective.

Over the years this series of **Understanding** books will provide a library of mini-tutorials on a variety of fundamental as well as technical subjects. Each book will serve as a rapid entry point or "short course" to a particular subject and we sincerely hope that the readers will reap immediate benefits when applying this knowledge to their research or work-related problems.

E.H. Immergut
Series Editor

Preface

Investigations of the form and structure, the morphology, of materials date back to very early times. Visual examination was followed by use of optical microscopy, of electron microscopy and most recently, of scanning probe microscopy, which allows imaging of individual molecules in the material. In the last half of this century there have been an incredible number of new materials prepared, many of which have been put to use in a variety of ways. Important among these materials are various synthetic and modified naturally occurring polymers and composites containing them. At present, both optical and electron microscopy are used routinely to study the morphologies of polymeric materials. Understanding and interpreting the varieties of morphology exhibited by polymeric systems is essential because of the direct effect the morphology has on properties and performance.

Various textbooks and monographs containing descriptions of some of the polymer morphologies known up to the 1980s have appeared. A relatively comprehensive pictorial presentation of the field was made in the recent *Atlas of Polymer Morphology* (Woodward, Hanser). However, a book designed to introduce and explain the various aspects of polymer morphology to scientists who are not specialists in this important field has not appeared to date. **Understanding** Polymer Morphology is an attempt to fill that need. It is believed that this book will be of use to many science and engineering students, as well as to those individuals preparing and studying polymeric materials but who are not yet specialists in this field.

In **Understanding** Polymer Morphology a brief introduction to pertinent polymer science principles, short descriptions of the microscopic methods used to study morphology and a discussion of crystallizable polymers are given. This book contains descriptions and molecular explanations of the various types of polymer morphologies reported to date, including those for crystallized systems, liquid crystalline mesophases, systems that phase separate, and those arising due to fabrication, deformation and fracture. Also discussed are: defects in polymer systems, imaging of individual molecules; specific molecular requirements for the appearance of particular morphologies; morphological changes brought about by various means; and the relationship of polymer properties to the morphology. A large number of the morphological investigations appearing in the literature, with an emphasis on more recent studies, are cited in Chapters 3 through 8. About 50% of the optical and electron micrographs given in the present book are from work reported after 1990. The reader is referred to the *Atlas of Polymer Morphology* for a collection of over 400 micrographs taken from work done prior to 1988 which illustrate many of the morphologies described herein.

I am most grateful to Professor Steven D. Hudson for a thorough and thoughtful review of the original manuscript. His many corrections and suggestions have been incorporated into an improved final draft.

Arthur E. Woodward
New York, New York

Contents

1   Introduction

1.1   Polymer Chain Structures

Polymer molecules, also known as *macromolecules*, are chains containing one or more *repeat units*, a particular grouping of atoms, linked together by covalent bonds. Polymers are important constituents of animals and plants; they can also be prepared in the laboratory by chemical reactions that cause the covalent linking of smaller molecules. Some examples of (partial) polymer chain structures are given below, using letters (A, B and C) to represent different types of chemical units.

```
linear homopolymer:     I-A-A-A-A......A-A-A-A-A-A-A-A-A-A-A-A-I'

branched homopolymer:  I-A-A-A-B-A-A-A-A......A-A-A-A-A-A-A-A-I'
                                 |
                                 A
                                 |
                       A-A-A-A-A......A-A-A-A-A-A-A-A-I'

polymer networks:   -A-A-A-B-A-A-A-A-A-B-A-A-A-A-A-A-A-A-
                          |               \
                          C                C
                           \               |
                       -A-A-A-B-A-A-A-A-A-B-A-A-A-B-A-A-
                                                 |
                                                 C
                                                 |
                                  -A-A-A-A-A-A-B-A-

and  -A-A-A-B-A-A-A-A-A-B-A-A-A-
            |           |
            A           A
            |           |
            A           A-A
            |           |
     -B-A-A-A-B-A-A-A-A-A-A-B-A-A-
      /            -A-A
     A              /
      \            /
       A-A-A-A-A-B-A-A-A-
```

statistical copolymer:
$$\text{I-A-A-A-A-B-A-A-A-B-A-A-B-A-A....A-A-A-B-B-A-A-A-I'}$$

block copolymer: I-A-A-A-A-A......A-A-A-A-A-A-A-A-A-B-B-B-B-B
$$\text{I'-A-A-A-A......A-A-A-A-A-B-B-B-B......B-B-B-B-B-B}$$

star block copolymer: I-B-B-B-B...B-B-B-A-A-A-A......A-A-A-A
$$\text{I-B-B-B-B...B-B-B-A-A-A-A......A-A-A-A-C}$$
$$\text{I-B-B-B-B...B-B-B-A-A-A-A......A-A-A-A}$$

In a *linear homopolymer* the chain contains the same chemical unit (A) repeated many times with two end-groups, I and I'. (I and I' may be the same or different.) If branching of the chain occurs, at least one unit per branch (represented by B above) differs from the other units (represented by A). Only one branch is shown in the illustration above, although more could occur. A crosslinked *network* can be formed in various ways. The first type shown above is a group of homopolymer chains linked together with a crosslinking agent, C; the second type shown involves the inclusion of a branching or crosslinking agent (B) in the polymerizing mixture. Linear homopolymers containing more than one type of repeat unit are *copolymers*. The copolymers, diagrammed above, contain two types of repeat units (A and B); copolymers can be prepared containing three or more different kinds of repeat units. A *statistical copolymer* contains sequences of repeat units that vary in length. The types of *block copolymers* illustrated above contain long sequences of two different repeat units (A and B) linked together. The *triblock copolymer* shown above, designated as an ABA type, contains a long sequence of A units, followed by a long sequence of B units, followed by a long sequence of A units. *Diblock copolymers* containing one block of A and one block of B units (AB type) and triblock copolymers containing long sequences of three different repeat units, A, B and C (ABC type), are also possible. The *star block copolymer* is a branched polymer in which the blocks of A are linked to a common multifunctional center, C. The number of branches shown in the star block copolymer example above is three; however, a larger number of branches is possible. A *graft copolymer* contains a chain of one type of repeat unit with branches made up of units of another type.

Polymer chains can contain any atom that will form two or more covalent bonds. To satisfy the valencies of the atoms linked together in the chain, additional atoms or groups of atoms are covalently attached. Synthetic polymers are prepared: 1) from unsaturated monomers, 2) by the reaction of functional groups containing monomers (difunctional or higher), or 3) from ring compounds. The polymers prepared from the latter two types of monomers are usually recognizable by the presence of functional groups, such as amide, carbonate, ester, ether, sulfide or urethane, in the chain. Many known polymer chains contain carbon atoms or aromatic structures covalently linked together by single bonds. Other atoms found in polymer chains covalently linked with carbon include: nitrogen, oxygen, phosphorus, silicon and sulfur. There are chains of atoms that do not contain carbon in the main chain, such as: 1) the phosphazenes, with alternating nitrogen and phosphorus atoms, 2) the siloxanes, containing alternating oxygen and silicon atoms, and 3) the silylenes, containing silicon atoms only.

1.2   Chain Configuration

Many repeat units derived from unsaturated monomers, such as that of $-CH_2CHR-$ in polystyrene in which R is a phenyl group, are asymmetric. Therefore, two adjacent units can be linked together in the following three ways to form different configurational isomers, as shown below using a generalized repeat unit, $-CX_2-CYZ-$:

head-to-tail (*isoregic*) $-CX_2-CYZ-CX_2-CYZ-$

head-to-head $-CX_2-CYZ-CYZ-CX_2-$

tail-to-tail $-CYZ-CX_2-CX_2-CYZ-$

(For polystyrene, the X and Y substituents are both H atoms.) Many, but not all, polymers with asymmetric repeat units have a predominately head-to-tail chain arrangement. Each of the above three arrangements can form additional geometric isomers because of the different possible placements of Y and Z. Considering a chain section containing three repeat units, three different arrangements are possible:

isotactic

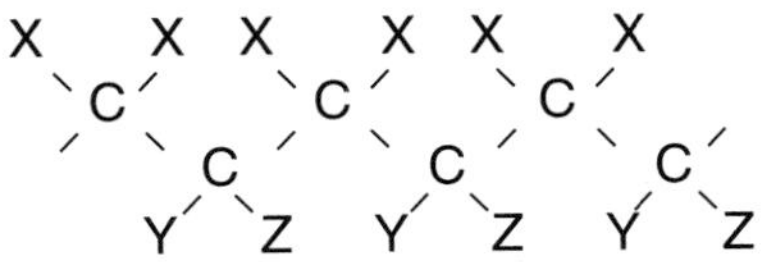

syndiotactic

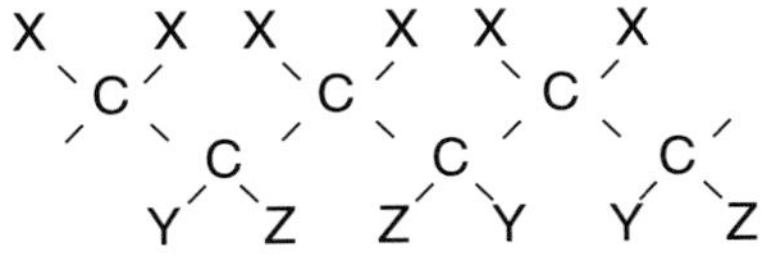

atactic

$$\begin{array}{ccccccc} X & & X & X & & X & X & & X \\ & C & & & C & & & C & \\ & & C & & & C & & & C \\ & Y & Z & Y & Z & Z & Y \end{array}$$

The four carbon atom orbitals in these chains are approximately tetrahedral; therefore, the Y-group projects in front or in back of the carbon atom plane and the Z-group does the opposite. *Isotactic* placement has the Y-group on either the front or the back in all three repeat units. In the *syndiotactic* configuration, the Y-groups alternate from front to back. The placement of Y-groups is mixed in the *atactic* configuration. Alkene units in the main chain, such as those in poly-1,4-butadiene or poly-1,4-isoprene, can form two different structural isomers:

cis

$$\begin{array}{cccc} H & H & H & H \\ & C & & C \\ & & C = C & \\ & R & & H \end{array}$$

trans

$$\begin{array}{cc} H & H \\ & C & H \\ & & C = C \\ R & & C \\ & & H & H \end{array}$$

1.3   Degree of Polymerization and Molecular Weight

The number of repeat units in a linear or branched polymer chain, the *degree of polymerization* (DP), is an important parameter in determining the properties of polymeric substances. The molecular weight is obtained from the DP upon multiplication by the repeat unit molecular weight and addition of the molecular weights of the end-groups. The DP for many linear or branched polymers is in the 100 to 10,000 range. Network polymers have an essentially infinite DP. A sample of a particular polymer containing chains, each with the same DP, is characterized as *monodisperse*. In most cases there is a distribution of chain lengths and an average DP is used; these samples are referred to as *polydisperse*. Different average DP's employed include the *number* average, DP_n, the *weight* average, DP_w and the *viscosity* average, DP_v. DP_n and DP_w are defined as follows: $DP_n = \Sigma N_i DP_i / \Sigma N_i$ and $DP_w = \Sigma N_i DP_i^2 / \Sigma N_i DP_i$, where N_i is the number of chains in the sample with DP_i. For a polydisperse sample, DP_w is larger than DP_n; the ratio DP_w / DP_n is used as a measure of the polydispersity. For coiled polymers, DP_v has a value between DP_n and DP_w. Certain properties, such as the viscosity, are dependent on the polymer DP over the total range available. Other properties, such as mechanical strength, show changes at low DP, then become constant, or nearly so, as the DP is increased to values above 100–1000.

1.4   Chain Conformation

Many polymers are composed of chains of singly bonded carbon atoms in tetrahedral or near tetrahedral bonding orbitals, as is shown for a portion of a polyethylene chain:

$$
\begin{array}{ccccccccc}
\text{H} & \text{H} & \text{H} & & \text{H} & \text{H} & & \text{H} \\
 & \diagdown \text{C} \diagup & & & \diagdown \text{C} \diagup & & & \diagdown \text{C} \diagup \\
 & \diagup & \diagdown & \text{C} & \diagup & \diagdown & \text{C} & \diagup & \diagdown \\
 & & \text{H} & \diagdown \text{H} & & \text{H} & \diagdown \text{H} & &
\end{array}
$$

In this illustration the four carbon atoms shown are in the same plane, and one of the hydrogen atoms on each carbon atom projects out and the other projects in. This is referred to as a trans conformation. The spatial placement of each CH_2 group and the remainder of the chain attached to it can be changed by rotation about the C–C bond between the two carbon atoms preceding the CH_2 in question. Each allowed change of the rotation angles yields another conformation for the chain. The chemical makeup of the polymer repeat units present dictates which conformations are energetically feasible as well as the relative energies of each. Rotation angles of 0°, ±60°, ±120° and 180° are labelled as trans (t), skew ± ($s^\pm$), gauche ± ($g^\pm$) and cis (c). For butane, a polyethylene precursor, the trans conformation has the lowest energy and the gauche ± has an energy that is 500 cal/mol higher than the trans. Although the trans is favored, the gauche ± states are also energetically feasible for butane at room temperature. The cis and skew ± are high energy states and are not occupied to a significant extent. For a polymer molecule, any combination of bond conformations that results in close approach or overlap of nonbonded atoms will be of high energy and, therefore, will not be preferred. For polyethylene, a succession of trans conformations (ttt) gives a chain with the lowest energy. However, many other combinations of trans, gauche + and gauche − occur for this polymer in the liquid state and in solution. The availability of nonextended conformations leads to the presence of (randomly) coiled chains. All linear and branched polymers with singly bonded chain units consist of randomly coiled molecules in the liquid state and in solution. The volume occupied by each molecule in the liquid state depends on the DP and the repeat units present. For example, this volume is larger for an atactic polystyrene chain than for a polyethylene chain with the same DP [1]. For isotactic polymer chains, a ttt conformation does not have the lowest energy; for the polydienes, with repeat units $-CH_2-CR=CH-CH_2-$, skew and cis states are energetically preferred to the trans and gauche states for the single bond following the double bond in the repeat unit. The presence of an inflexible chain part, such as a terephthalate unit,

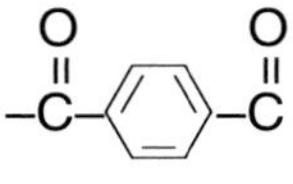

Limits the conformations available. The all-aromatic polyamides and polyesters are semi-rigid with the following conformational choices available:

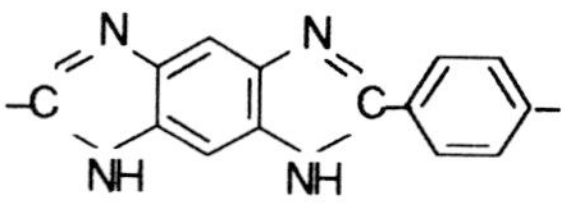

where R is the remainder of the chain. Rotation about the $\overset{O}{\overset{\|}{C}}-NH$ bond takes considerable energy, resulting in a fixed conformation for that group. A rod-like polymer, such as polybenzo-imidazole with the repeat unit:

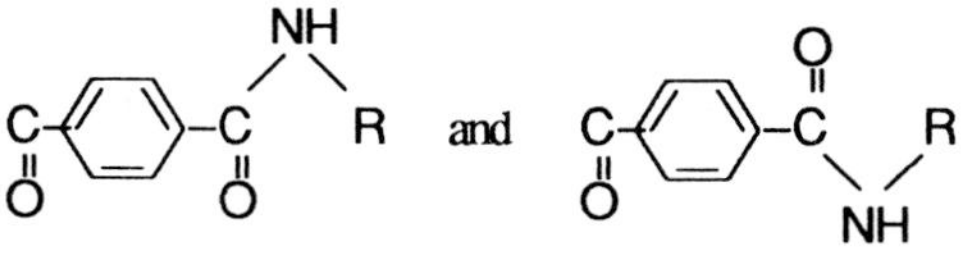

has only one conformation available to it and is fully extended.

1.5   Molecular Forces

Separate molecules are held together in condensed phases (i.e. isotropic liquids, liquid crystalline mesophases and crystalline solids), by van der Waals forces and by hydrogen bonding. The type and strength of these intermolecular interactions depend on the chemical repeat unit(s) incorporated in the polymer chains. Van der Waals forces include dipole/dipole interactions, the attractive interaction between the positive end of a dipolar grouping of atoms in a molecule for the negative end of another dipolar grouping on a nearby portion of the same or another molecule. If hydrogen linked to a highly electronegative atom, such as oxygen or nitrogen, is part of one of the dipolar groupings involved, then this type of attraction is usually referred to as hydrogen bonding. The presence of a dipole can induce polarization of other atomic groupings in nearby parts of the same or a different molecule. Attraction will occur between oppositely charged parts of the two; this is called dipole/induced dipole attraction. All atoms, regardless of their polarity, have fluctuations occurring in their electronic clouds about the atoms, leading to attractive interaction between neighboring nonbonded atoms. These are called London dispersion forces. All van der Waals forces of attraction are inverse functions of distance.

1.6   Polymer Liquids

Due to their long chain nature, polymer liquids have high viscosities when compared to small molecular liquids. The presence of chain entanglements accounts for the change in molecular weight dependence of the melt viscosity from M to M^3 for various polymers, occurring at a DP of about 1000 [2]. Due to this change, processing of polymer melts becomes more difficult with increasing molecular weight above DP = 1000.

Linear and branched polymer molecules, composed of small flexible units, show no long-range molecular order in the liquid state; they are *isotropic*. This lack of long-range order exists even at temperatures where the sample is *glassy* (hard and brittle), as well as at temperatures at which it is a *rubber* (soft and extensible). Changes in conformation of each chain in a rubber occur with time, but the average conformation remains the same if the system is contained or is crosslinked. Unless a crosslinked rubber is subjected to an applied stress, the overall sample dimensions do not change. In an unstressed glassy liquid each chain has a particular conformation, and conformational changes do not occur with time.

Some types of polymer molecules can exhibit one- or two-dimensional order in the liquid state over a limited temperature range, due to partial main- or side-chain alignment associated with rigid portions of the chain; this behavior is characteristic of *thermotropic liquid crystalline mesophases*. Partial order has also been observed in some solutions of particular polymers and are referred to as *lyotropic* liquid crystalline mesophases. The appearance of lyotropic liquid crystalline behavior is both concentration and temperature dependent.

1.7   Crystallization

For many polymers, the chains, or portions thereof, can arrange themselves in a three-dimensional repeating array of atoms. Each separate group of repeat units in a crystallite assumes a particular conformation and is packed in a particular arrangement with respect to the other groups present. For polymers with flexible chains, where more than one conformation for successive units exists, the portions of the chains present in the crystal lattice must all adopt the same low energy conformation. For some polymers, this is the most extended conformation possible, but for many polymers the conformation adopted is a helical one with either an odd, an even, or a fractional number of chain repeat units per turn of the helix. The crystallographic repeating unit, the unit cell, in a polymer crystal lattice usually contains small portions of a few chains. For example, the orthorhombic unit cell in polyethylene contains four CH_2 groups on two all-trans chains, and the triclinic unit cell in polytetrafluoroethylene contains $13\,CF_2$ units on one helical chain. The crystallite dimensions are many times larger than those of the unit cell and, therefore, each chain section in a crystallite passes through many unit cells. Many polymers can crystallize in more than one arrangement. These different crystal lattices for the same polymer may involve different chain conformations, a different number of chains, or both. Two different crystal lattices can exist for the same polymer at the same temperature and pressure or can occupy separate temperature/pressure regions.

Polymers can be crystallized by cooling the rubbery liquid to a temperature below the equilibrium melting point but above the temperature at which the polymer exists as a glass. Prior to the onset of crystallization, the liquid exists in a *supercooled* condition. Crystallization can also be brought about by supercooling a solution of the polymer or by evaporation of the solvent.

Polymers crystallize as single lamellar platelets, as multilamellar platelet and ribbon structures, as microfibrils, or as a combination of platelets and fibrils. Most of these structures contain noncrystallized portions of chains which are associated with the liquid state. This type of structure is referred to as *microcrystalline* or *paracrystalline*.

Discussion of the types of polymers that crystallize and the effects of structure on the melting process are given in the next chapter.

1.8   Processing Methods

The methods used in the processing of polymer films include *extrusion*, *calendering* and *(bubble) blowing*; tubes are made by extrusion. In the extrusion process, melted polymer is pushed through a slit or orifice. Calendering involves pressing of a polymer melt between heated rollers. Bubble blowing includes extruding a cylinder, expanding this using compressed gas, and flattening between rollers [3].

Fibers can be produced by pushing (extrusion of) a melt through a small hole into a region kept at a lower temperature. Fibers can be formed by extrusion of a solution into a precipitant or into a heated chamber in which rapid evaporation of the solvent occurs. Yarns and woven fabrics are formed from oriented fibers. Nonwoven fabrics can be prepared by spray spinning and by calendering [4].

For the fabrication of shaped objects various molding techniques are used; these include *compression, injection, blow, rotational* and *bag molding*, thermofusion and thermoforming. The morphology that results can vary considerably from method to method due to marked differences in processing conditions.

Fabricated samples for commercial use can contain one or more polymer components; other materials such as antioxidants, crosslinking agents, fillers, flame retardants, foaming agents, lubricants, plasticizers and ultraviolet light absorbers may be present in a formulation. Blends of two or more polymers can be prepared by mixing in the melt or in solution. Composites contain a fibrous or particulate solid mixed or layered with a thermoplastic or a thermosetting polymer.

1.9   Thermal Transformations

The physical state of a polymeric material depends on the material, the temperature and the previous thermal and mechanical stress history. The phases that can be found for a flexible chain polymer, in order of decreasing temperature are: isotropic (rubbery) liquid > liquid crystalline rubber mesophase > microcrystalline rubber > glassy liquid, liquid crystalline glass and microcrystalline glass. Upon heating a polymer, various changes can occur at constant or near constant temperature and are accompanied by the evolution of heat and a change in volume; these are considered to be first-order processes and have particular transformation temperatures associated with them. These processes include: 1) a change from one crystal form to another, 2) transformation of a microcrystalline rubber to a liquid crystalline rubber mesophase or to an isotropic rubber phase, and 3) change of a liquid crystalline mesophase to an isotropic rubber liquid phase. The change of a microcrystalline glass to a microcrystalline rubber, a liquid crystalline glass to a liquid crystalline rubber, or a isotropic glass to a rubbery liquid are characterized by a glass transition temperature, T_g; these are not first-order transitions and are affected by kinetic factors.

Highly crosslinked networks are usually glassy and show few changes with changing temperature. Lightly crosslinked networks can crystallize and show rubber–glass changes.

Heating a polymer to moderate or high temperatures, usually in the presence of oxygen, can cause chemical changes with breaking of the chain or degradation occurring. When polymer materials are brought in contact with interactive liquids or vapors, plasmas and irradiation, changes can occur due to dissolution, crystallization or degradation.

1.10   Mechanical Stress Effects

The types of *stress* (force/unit area) that can be applied to an object are *tensile*, *shear* and *compressive*. The change in specimen dimensions which occurs depends on the stress applied, as shown in Fig. 1.1. When a tensile stress is applied, the sample dimensions increase in the drawing direction and decrease

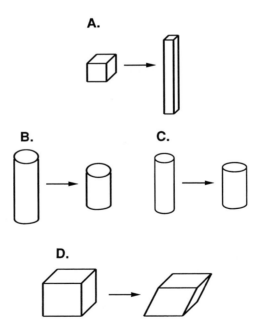

Figure 1.1   Types of mechanical deformation: A. tensile, B. plane strain compressive, C. uniaxial compressive, and D. shear.

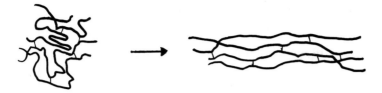

Figure 1.2   Conformation changes during elongation of a crosslinked rubber.

in the directions at right angles to the applied stress. The *strain* (γ) resulting from a tensile stress is equal to the change in length divided by the rest length of the sample and can be given as a percentage. Another parameter used to express the tensile strain is the extension ratio (α), which is equal to $1+\gamma$. Two types of compression are illustrated in Fig. 1.1, *plane strain* and *uniaxial*. Plane strain compression is carried out on a sample that is confined in a metal die, and only the dimension in the compression direction changes. In uniaxial compression, the dimension in the compression direction decreases and the dimensions at right angles to this direction increase. The magnitude of the dimensional changes that take place upon application of a stress to a polymer sample depends also on the state of the sample and can vary widely.

The alterations in chain conformation that occur upon application and removal of a tensile stress to a rubbery network are shown in Fig. 1.2. The application of an elastic stress to a rubbery material leads to relatively large extensions and the adoption of extended chain conformations, as shown. If the chains are not linked together by covalent bonds or by physical forces into a network, they will return to an equilibrium arrangement of segments by a viscous flow process, leading to a change in dimensions for the sample. If all the chains in the specimen are linked by covalent bonds, removal of the stress will lead to a return of the sample to its original dimensions, assuming no chain scission has occurred in the stressed state.

Application of mechanical stress to a glassy polymer leads only to small changes in sample dimensions prior to fracture. A large portion of the change in dimensions that occurs is not recoverable upon removal of the stress.

Microcrystalline lamellar structures can show large degrees of *yielding*, dimensional changes with little or no increase in stress, and *permanent deformation* upon stress application. This process leads to *fibril* formation.

The *tensile strength* of a material is the tensile stress necessary to cause fracture. This parameter depends on the temperature and the rate of stress application. The stress necessary to cause yielding is termed the *yield stress. Impact strength* is the stress applied in an impact test that causes fracture.

1.11 Phase Separation in Noncrystallizing Systems

Most mixtures of two nonpolar polymers at equilibrium in the liquid state will separate into domains predominantly containing the individual components. Phase separation happens because the forces of attraction between nonpolar molecules are relatively weak, leading to a positive enthalpy of mixing, ΔH_m; energy is taken up in the mixing process. This unfavorable effect is opposed by the increase in randomness of the system that occurs on mixing, leading to a positive entropy of mixing, ΔS_m. Incompatibility is characterized by a positive free energy of mixing, ΔG_m; at equilibrium: $\Delta G_m = \Delta H_m - T\Delta S_m$, where T is the temperature (in $^\circ$K). The disposition and size of the domains observed for phase-separated polymer blends depends on the method of sample preparation. Favorable interactions between two dissimilar polymer chains containing polar groups can yield a compatible system giving only one phase.

As pointed out above, block copolymers are made up of sequences of two or more repeat units. Microphase separation of the components of a nonpolar block copolymer can take place.

1.12 Methods of Morphological Investigation

The supramolecular structure of polymeric systems can be viewed by optical and electron microscopy [4, 5]. Lattices can be imaged and electron diffraction patterns of crystalline systems obtained using the electron microscope. Molecular structure can be revealed using a scanning probe microscope [6].

1.12.1 Optical Microscopy

Optical microscopy is carried out on solids and solid–liquid mixtures at magnifications up to about 2000×. A transmitted light beam that is *plane polarized* or is subjected to *phase contrast* or to *interference contrast* can be employed.

Viewing of images formed with plane polarized light is done through an analyser, a polaroid lens with its axis usually oriented at right angles to that of the polarizer. Light can be transmitted through the analyser if part or all of the sample is *birefringent*, that is, has a refractive index for light with electric vibrations parallel to the optic axis different from that for light perpendicular to this axis. The polarization of light passing through a birefringent material is usually changed, and the birefringence can be a function of the wave length.

Even in birefringent materials, light will not be transmitted; that is, optical extinction will occur for the following reasons: 1) the thickness of the birefringent area is less than one wavelength, 2) the optical axis is parallel to the polarizer or analyser direction, or 3) the direction of the light is parallel, and therefore the electric vector is perpendicular, to the optic axis.

Phase contrast uses a 90° change in phase of the light diffracted from various regions of the sample; this shows differences in refractive index within the specimen when recombined with the transmitted beam.

In interference contrast, one portion of the beam passes through the sample and the other part is changed in phase by a variable amount to give optimum contrast upon recombination of the two beams, showing steps and surface tilts present and giving a three-dimensional view.

Rotation of polarization by the medium is classed as a *second-order nonlinear optic* (NLO) phenomenon [7]. Other second-order NLO effects are: 1) the combination of two photons to form a new photon, 2) the subtraction of two photons, and 3) frequency and amplitude modulation. Third-order NLO effects are observed for some materials. NLO effects in polymers can be studied using a high intensity monochromatic light source, as supplied by a laser.

1.12.2   Electron Microscopy

Electron microscopy uses an electron beam to investigate the sample. This can be carried out in transmission on relatively thin polymer specimens at magnifications up to about $2 \times 10^6 \times$. *Transmission electron microscopy* (TEM) is usually carried out in *bright field*, the transmitted electrons being used to form the image. It is possible to reverse the contrast by forming the image in *dark field* with one or more of the diffracted beams. Dark field techniques are used to investigate dislocations in crystalline polymers with *Moiré patterns* being produced under certain conditions. If the sample is crystalline, a *selected area electron diffraction* pattern can also be obtained. Phase contrast in the transmission electron microscope is obtained by defocusing the objective lens, which produces a phase shift. This method can be used for the detailed study of polymeric order. However, artifacts, particularly in randomly structured specimens, can be introduced by defocus [4].

Samples for bright field TEM examination are usually shadowed with a heavy metal to provide contrast between the specimen and the carbon film supporting it on the sample grid.

Polymer samples usually undergo a high degree of crosslinking and/or degradation at normal beam currents in a transmission electron microscope, leading to a significant crystallinity loss. However, by using a *high resolution electron microscope* (HREM) at low dosages and moderate magnifications, followed by photographic enlargement and computer-controlled image en hancement, a crystalline or liquid crystalline lattice can be imaged.

Another type of microscopy, *scanning electron microscopy* (SEM), uses a focused electron beam to scan the sample surface. Maximum resolution is obtained with a high accelerating voltage, small probe size and high beam current. These conditions may result in beam damage to sensitive specimens. The depth of focus decreases with increasing magnification and is about 1 μm at 10,000× magnification. Magnifications up to about 100,000× are possible, although most work is carried out below 10,000×. In addition to having a higher resolution than optical microscopy, SEM has a much larger depth of field. Samples are thinly coated with a metal to provide a conductive layer.

1.12.3   Scanning Probe Microscopy

The most recent development in microscopy are the scanning probe microscopes [6, 8]. These involve moving a probe with a finely ground tip from one to a few atoms across, as shown in Fig. 1.3, over the surface of the specimen. The specimen surface can be imaged with a small electrical current, as used in the *scanning tunneling microscope* (STM), or by application of a mechanical force, as employed in the *atomic force microscope* (AFM). The STM uses a tungsten probe and requires a conducting surface; the AFM employs a diamond probe and can be used to investigate insulative surfaces, including polymers. In both the

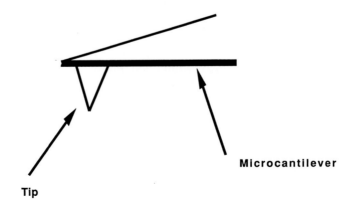

Figure 1.3   Atomic force microscope probe.

STM and the AFM, the electron cloud at the tip of the probe is in contact with that at the specimen surface. With the use of the STM and AFM, a topographical investigation is carried out on an atomic scale. Atom-resolved images of polymer molecules have been obtained. Scanning probe microscopes that use electrostatic or magnetic forces and do not damage or contaminate the sample surface have been developed [6]. The components of a force microscope include: 1) a sharp tip attached to a weak cantilever spring (Fig. 1.3), 2) a spring deflection sensor, 3) a feedback system, 4) a scanning system, and 5) a display system [8].

1.12.4   Sample Preparation and Treatment

Many polymer samples are too thick for TEM investigation using conventional microscopes with accelerating voltages of 200 kV. Microtoming sections 200 nm or less in thickness or *replication* of brittle fracture or peeled back surfaces are two preparative methods used to allow investigation of such samples. To replicate, the surface is coated with a polymer solution, the solvent evaporated, the dried film stripped off, and the sample side coated with evaporated carbon. After the carbon side is coated with evaporated heavy metal, the replicating film is dissolved. The coated carbon replica is then placed in the microscope. The solvent used in the replication process must have no effect on the surface being copied.

If a higher accelerating voltage than the conventional 200 kV is used, thicker samples can be explored without replication. Samples as thick as 5 μm have been investigated at an accelerating voltage of 1000 kV [9]. Prior to replication or following microtoming, various chemical treatments can be carried out to highlight or reveal certain parts of the total solid sample. Reaction with OsO_4 is particularly useful for homopolymers, block co-polymers and blends containing unsaturated hydrocarbon repeat units. Attachment of osmium atoms to the chains results in a darker image for the sample portions containing these repeat units. RuO_4 is used in a similar fashion with polymers such as polystyrene, the polyamides and polyethylene. Etching with acids, bases, organic liquids, ions or electrons has been employed to strip away portions of the sample.

Features on polymer surfaces can be enhanced by decoration using either the same material in a different form, referred to as self-decoration, or a different material. One type of decoration involves the use of heavy metal atoms; however, heavy metal decoration is not a process in which the complete surface is covered, as in shadowing of polymer samples. Self-decoration of a polymer surface implies the use of polymer molecules, not single atoms, in the decoration process.

1.13   Importance of Morphology

The physical properties and the response to an applied mechanical stress exhibited by a material can depend markedly on the supramolecular structure or morphology of the sample. The morphology depends on the components present, the fabrication method(s) used in the sample preparation, and the thermal history. For example, consider a system composed of a flexible chain polymer that crystallizes in a lamellar habit. Crystallization can be carried out by a variety of means which include: 1) precipitation from a quiescent solution, 2) precipitation from a stirred solution, 3) crystallization of a film from the melt, or 4) melt crystallization of a film under an applied tensile force. The first type of crystallization, after removal of the liquid, yields a partially crystalline powder; the second procedure, after liquid removal, gives a precipitated fibrous mass; the third method results in a film containing randomly oriented crystallites connected by portions of chains; and the fourth procedure gives a film with a common chain direction. The first of these is made up of individual multilamellar structures with little or no coherence to each other; the second will be partially coherent but will have no form; the third will contain interlinked multilamellar structures but little or no lamellar orientation will be present; and the fourth will have a large amount of chain orientation in fibrils and lamellas in the direction in which the force was applied. These samples would show completely different responses to a tensile load, and their physical properties, such as crystallinity and density, would be different.

Both the response to an applied stress and the physical properties of a polymer system will depend on: 1) the amount of chain extension and mobility, 2) the amount of phase separation in a blend or a block copolymer, 3) the domain order in a phase-separated blend or block copolymer, and 4) the sample homogeneity. The response to an applied stress will also depend on: 1) the direction of chain orientation, 2) the domain size in a phase-separated system, 3) arrangement of components in a composite material, and 4) sample continuity. Changes in morphology, brought about by an interactive environment or by a change in temperature, will cause changes in physical properties and in the response to an applied stress.

1.14   Polymer Morphologies

Distinctive morphologies are observed for: 1) crystallized polymers, 2) polymers in liquid crystalline mesophases, and 3) block copolymers containing noncrystallizable and incompatible chain sections. Liquid crystalline mesophases show arrangements and textures that are dependent on the molecular structure of the polymer and experimental parameters; crystallized polymers can have various morphologies, depending on the crystallization conditions and postcrystallization treatment, and the morphologies exhibited by noncrystallizable block copolymers are a function of the block types, the component molecular weights, and fabrication conditions. Since more than one type of morphology can be adopted by polymers in categories 1, 2 and 3, the distinctions are numer-ous.

Many of the distinct morphological features described in this book are the result of physical processes, such as crystallization, phase separation, flow, deformation and fracture. The morphologies arising due to liquid phase separation depend on the chemical composition and, to some extent, the molecular weight of the components. Whether a polymer is crystallizable depends primarily on the chain configuration. The amount of crystallization taking place at a particular temperature is a function of the chemical composition, the molecular weight, molecular weight distribution and the processing conditions. Flow, deformation and fracture morphologies are affected by chain configuration, chemical composition, molecular weight and molecular weight distribution.

Insights into and information about structural details of polymer crystal lattices were based originally on X-ray diffraction studies of fibers and electron diffraction investigations of lamellas. More recently, imaging of lattices in lamellar crystals, microfibrillar crystals and glassy liquid crystalline mesophases has been accomplished using magnified TEM photos that confirm diffraction investigations on similar specimens. The newer imaging techniques, STM and AFM, have been used to study the surfaces of various polymer preparations, providing micrographs at magnifications that allow resolution of the individual chains.

In Chapter 2, the various types of crystallizable homopolymers will be introduced and some of their properties discussed. The recent role of surface imaging to view polymer molecules in their crystal lattices will be reviewed. In Chapter 3, the morphologies observed for crystallized polymers, including single lamellas, branched structures and multilamellar arrays, produced under quiescent conditions, and the fibrous and epitaxial structures arising in the presence of an applied stress, will be described and conditions for their appearance enumerated. The importance of various types of defects and the multiphase aspect of polymer crystallization will also be discussed. A discussion of the types of liquid crystalline mesophases that occur, the texture of these mesophases and their imaging will be given in Chapter 4. The prevalence of incompatibility in polymer mixtures and descriptions of morphologies associated with phase-separated polymer blends, block copolymers and composites will be addressed in Chapter 5. The multiple morphologies that occur when molding and other forms of processing are employed, as well as some of the effects of environmental exposure, are described in Chapter 6. The effects of mechanical stress on polymers present as rubbers, glasses or partially crystalline materials, resulting in permanent deformations and fracture, are described in Chapter 7. Finally, the relationships between morphology and the properties related to crystallinity and the mechanical, optical and electrical behavior are explored in Chapter 8.

References

1. Flory, P.J. (1988) *Statistical Mechanics of Chain Molecules*. Munich: Hanser Publishers.
2. Fox, T.G. and Allen, V.R. (1964) J. Chem Phys. *41*, 344.
3. Allcock, H and Lampe, F. (1990) *Contemporary Polymer Chemistry*, 2nd. Ed. Englewood Cliffs, NJ: Prentice-Hall, Inc.
4. Sawyer, L.C and Grubb, D.T. (1987) *Polymer Microscopy*. London: Chapman and Hall.
5. Woodward, A.E. (1988) *Atlas of Polymer Morphology*. Munich: Hanser Publishers.
6. Wickramasinghe, H.K. (1989) Sci. Am., *271* (4), 98.
7. Sperling, L.H. (1992) *Introduction to Physical Polymer Science*, 2nd Ed. New York: WileyInterscience.
8. Rugar, D. and Hansma, P. (1990) Physics Today, (Oct.), 23.
9. Michler, G.H. and Gruber, K. (1976) Plaste u. Kautschuk *23*, 496.

2   Crystallizable Polymers

Very definite differences in morphology occur between flexible chain polymers that crystallize to a significant extent and those that do not. The ability of a polymer to form a three-dimensional repeating structure is largely, but not totally, dependent on the regularity of the atomic placement in the chain. Homopolymers that crystallize can be obtained by polymerization of unsaturated monomers, by condensation-type reactions of bifunctional monomers, and by polymerization of certain ring monomers.

2.1   Polymers from Unsaturated Monomers

Repeat unit sequences for some crystallizable homopolymers from unsaturated monomers, such as propylene: $CH_2=CH(CH_3)$, are given in Table 2.1; T_g and T_m values are also shown. Some of the repeat units that form singly bonded carbon atom chains are symmetrical, having the same substituents attached to each carbon atom in the chain, e.g., the repeat unit in polyethylene; some have different types of substituents on alternate carbon atoms but have isoregic linking of the repeat units. For the latter the substituents on a given carbon atom can be the same, as for poly(vinylidene chloride) or they can be different, e.g., polypropylene, and are referred to as asymmetric. The polymers with asymmetric repeat units that crystallize usually have a regular configuration, isotactic or syndiotactic. The polymers with asymmetric repeat units in an atactic configuration that crystallize are those with small highly polar substituents such as the CN group in poly(acrylonitrile) and the OH group in poly(vinyl alcohol). Atactic poly(methyl methacrylate), atactic polystyrene and many other homopolymers from unsaturated monomers do not crystallize. These polymers contain relatively large asymmetric repeat units which are irregular in their stereochemistry or configuration with respect to one another. An example of a polymer with an asymmetric repeat unit, shown by NMR analysis to be nonisoregic as well as atactic, is free radical initiated poly(vinyl fluoride) [1]; despite these irregularities,

Table 2.1   Crystallizable Polymers from Unsaturated Monomers[a]

Name	Repeat unit sequence	T_g °C	T_m °C
polyethylene	$-\overset{\overset{\displaystyle H}{\vert}}{\underset{\underset{\displaystyle H}{\vert}}{C}}-\overset{\overset{\displaystyle H}{\vert}}{\underset{\underset{\displaystyle H}{\vert}}{C}}-$	−125	140
polytetrafluoroethylene	$-\overset{F}{\underset{F}{C}}-\overset{F}{\underset{F}{C}}-$	−113, 130	30, 327
poly(vinylidenechloride)	$-\overset{Cl}{\underset{H}{C}}-\overset{Cl}{\underset{H}{C}}-\overset{Cl}{\underset{H}{C}}-\overset{Cl}{\underset{H}{C}}-$	−18	210
poly(vinylidenefluoride)	$-\overset{F}{\underset{H}{C}}-\overset{F}{\underset{H}{C}}-\overset{F}{\underset{H}{C}}-\overset{F}{\underset{H}{C}}-$	−39	171
isotactic polypropylene	$-\overset{H}{\underset{H}{C}}-\overset{CH_3}{\underset{H}{C}}-\overset{H}{\underset{H}{C}}-\overset{CH_3}{\underset{H}{C}}-\overset{H}{\underset{H}{C}}-\overset{CH_3}{\underset{H}{C}}-$	26	150
syndiotactic polypropylene	$-\overset{H}{\underset{H}{C}}-\overset{CH_3}{\underset{H}{C}}-\overset{CH_3}{\underset{H}{C}}-\overset{H}{\underset{H}{C}}-\overset{H}{\underset{H}{C}}-\overset{CH_3}{\underset{H}{C}}-$	—	138
isotactic polystyrene $(R = \text{phenyl})$	$-\overset{H}{\underset{H}{C}}-\overset{R}{\underset{H}{C}}-\overset{H}{\underset{H}{C}}-\overset{R}{\underset{H}{C}}-\overset{H}{\underset{H}{C}}-\overset{R}{\underset{H}{C}}-$	100	240
isotactic polybutene-1	same as isotactic polystyrene but $R = CH_2 - CH_3$	−24	138[b]
isotactic poly(4-methylpentene-1)	same as isotactic polystyrene but $R = CH_2 - CH(CH_3)_2$	40[c]	250[b]
isotactic poly(methyl vinyl ether)	same as isotactic polystyrene but $R=OCH_3$	−21	150

Table 2.1 (continued)

Name	Repeat unit sequence	T_g °C	T_m °C
isotactic poly(methyl methacrylate) (R =– C– O – CH₃)	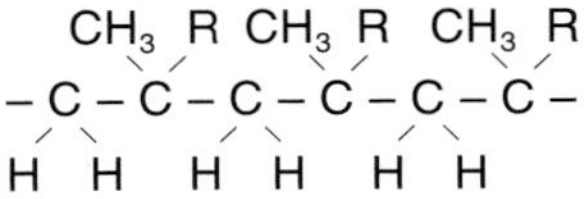 (structure)	48	160
syndiotactic poly(methyl methacrylate)	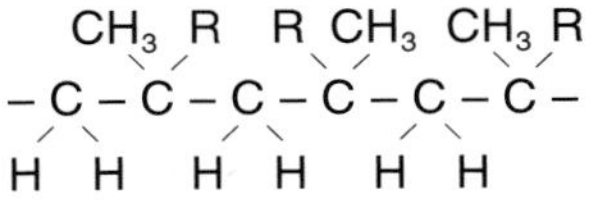 (structure)	128	200
trans-1,4- polybutadiene	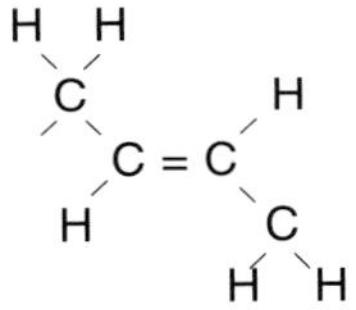 (structure)	–85	67, 133[d]
trans-1,4- polyisoprene	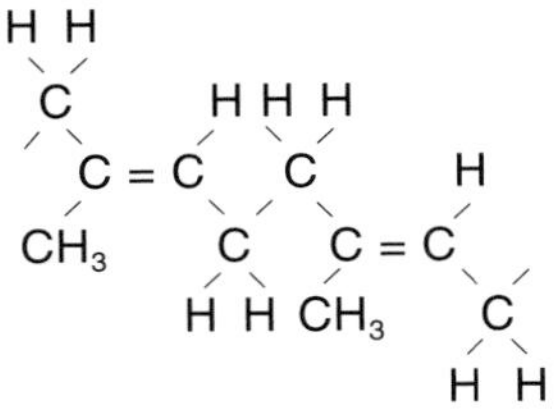 (structure)	–68	80[e]
N-carbazole substituted polydiacetylene (R = CH₂NC₁₂H₈)	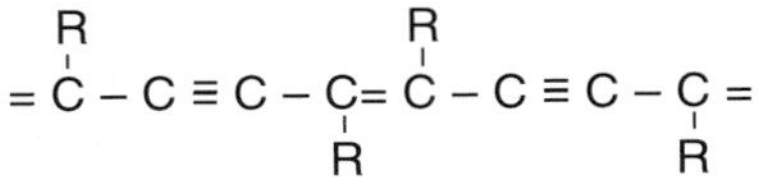 (structure)	—	– 214[f]
atactic polyacrylonitrile	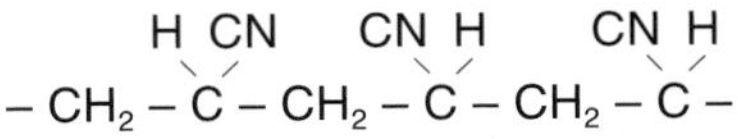 (structure)	85	317
atactic poly(vinyl alcohol)	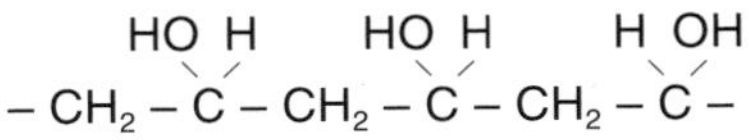 (structure)	99	258
atactic polytrifluorochloroethylene	same as atactic poly(vinyl alcohol) substituting F for H and Cl for OH	–40, 70[c]	295[g]
poly(vinyl fluoride)	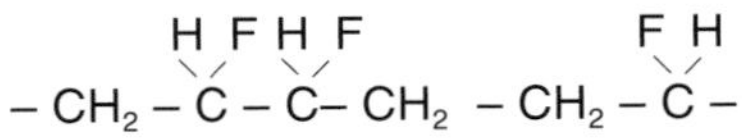 (structure)	30[c]	200

isotactic poly(methyl methacrylate) $(R =- C- O - CH_3)$

$$
\begin{array}{cccccc}
CH_3 & R & CH_3 & R & CH_3 & R \\
-C & -C & -C & -C & -C & -C- \\
H & H & H & H & H & H
\end{array}
$$

syndiotactic poly(methyl methacrylate)

$$
\begin{array}{cccccc}
CH_3 & R & R & CH_3 & CH_3 & R \\
-C & -C & -C & -C & -C & -C- \\
H & H & H & H & H & H
\end{array}
$$

atactic polyacrylonitrile

$$
\begin{array}{ccccccc}
& H & CN & & CN & H & & CN & H \\
-CH_2 & -C & -CH_2 & -C & -CH_2 & -C-
\end{array}
$$

atactic poly(vinyl alcohol)

$$
\begin{array}{ccccccc}
& HO & H & & HO & H & & H & OH \\
-CH_2 & -C & -CH_2 & -C & -CH_2 & -C-
\end{array}
$$

poly(vinyl fluoride)

$$
\begin{array}{cccccc}
H & F & H & F & & F & H \\
-CH_2 & -C & -C- & CH_2 & -CH_2 & -C-
\end{array}
$$

[a] All values of T_g and T_m were taken from Ref. [8] unless noted otherwise; [b] Ref. [2]; [c] Ref. [3]; [d] Ref. [4]; [e] Ref. [5]; [f] Ref. [6]; [g] Ref. [7].

this polymer crystallizes. Polymers synthesized from diene monomers are crystallizable if the linkage is 1,4 and if the substitution is predominately cis or trans, or if an isotactic or syndiotactic arrangement occurs in the 1,2 or 3,4 substituted isomers.

Statistical copolymers containing a symmetrical repeat unit and small amounts of a second unit, such as copolymers of ethylene with small amounts of propylene or vinyl acetate, are crystallizable; the repeat unit present in small amounts may be excluded from the crystal stems and be included in a non-crystallizing surface portion. The rejection of the noncrystallizing units, when it occurs, causes additional crystallizable units to be rejected as well [9]. Statistical copolymers of two or more asymmetric monomers usually do not crystallize.

The chains formed from unsaturated monomers are generally flexible due to the presence of many singly bonded carbon atoms. The effects of this flexibility on the morphology will be discussed in Chapter 3.

2.2   Polymers from Difunctional Monomers

Many polymers are synthesized from monomers with reactive end- or side-groups, such as alcohols, acids or amines. Homopolymers formed from difunctional monomers are usually crystallizable since only one isomeric choice exists and ordered arrangements can be formed. The repeat units of various crystallizable polymers containing main-chain functional groups are listed in Table 2.2. Values for T_g, T_m and an additional transition temperature, T_i, are included in this table. Discussion of the melting behavior of these polymers is discussed below. Many of the polymers listed in the table contain unsubstituted aromatic rings linked in the 1,4 position. The introduction of 1,2 and 1,3 linkages will cause disorder in the crystal lattice and thereby lower the crystallization ability. Substituents on the rings will have a similar effect on the crystal lattice. The repeat units shown in Table 2.2 include flexible chains, e.g., poly(hexamethylene adipamide) and poly(ethylene terephthalate); semiflexible chains, such as hydroxypropyl cellulose, polycarbonate, poly(ethylene terphenylate), poly(p-phenylene oxide), poly(p-phenylene sulfide); semi-rigid chains, such as the totally aromatic polyamides, and rigid chains, such as the polybenzazoles. These different types of polymers show different crystallization characteristics, particularly in their ease of crystallization and, to some extent, in the morphologies that appear.

The presence of tri- or tetra-functional monomers in a condensation-type reaction mixture leads to highly branched or to network polymers which usually

Table 2.2   Crystallizable Condensation-Type Polymers[a]

Name	Repeat unit	T_g (°C)	T_m (°C)	T_i (°C)
poly(hexamethylene adipamide) (nylon 66)	$-\text{NH}-(\text{CH}_2)_6-\text{NH}\overset{\text{O}}{\overset{\|}{\text{C}}}-(\text{CH}_2)_4-\overset{\text{O}}{\overset{\|}{\text{C}}}-$	45	267	—
poly(11-amino-undecanoic acid)	$-\text{NH}(\text{CH}_2)_{11}-\overset{\text{O}}{\overset{\|}{\text{C}}}-$	46	198	—
poly(nonamethylene urea)	$-\text{NH}(\text{CH}_2)_9-\text{NH}-\overset{\text{O}}{\overset{\|}{\text{C}}}-$	—	236	—
poly(trimethylene urethane)	$-(\text{CH}_2)_3-\text{O}-\text{NH}-\overset{\text{O}}{\overset{\|}{\text{C}}}-$	—	148	—
poly(ethylene terephthalate)	$-\text{O}-(\text{CH}_2)_2-\text{O}-\overset{\text{O}}{\overset{\|}{\text{C}}}-\langle\!\bigcirc\!\rangle-\overset{\text{O}}{\overset{\|}{\text{C}}}-$	69	270	—
poly(m-phenylene adipamide)	$\overset{-\text{NH}}{\langle\!\bigcirc\!\rangle}-\text{NH}-\overset{\text{O}}{\overset{\|}{\text{C}}}-(\text{CH}_2)_4-\overset{\text{O}}{\overset{\|}{\text{C}}}-$	—	296[b]	344[b]
polycarbonate	$-\text{O}-\langle\!\bigcirc\!\rangle-\overset{\text{CH}_3}{\underset{\text{CH}_3}{\text{C}}}-\langle\!\bigcirc\!\rangle-\text{O}-\overset{\text{O}}{\overset{\|}{\text{C}}}-$	150	267	—
poly(p-phenylene oxide)	$-\overset{\text{CH}_3}{\underset{\text{CH}_3}{\langle\!\bigcirc\!\rangle}}-\text{O}-$	—	338	—
poly(p-phenylene sulfide)	$-\langle\!\bigcirc\!\rangle-\text{S}-$	85	288	—
polybenzamide	$-\text{NH}-\langle\!\bigcirc\!\rangle-\overset{\text{O}}{\overset{\|}{\text{C}}}-$	—	lyotropic nematic	
poly(p-phenylene terephthalamide)	$-\text{NH}-\langle\!\bigcirc\!\rangle-\text{NH}\overset{\text{O}}{\overset{\|}{\text{C}}}-\langle\!\bigcirc\!\rangle-\overset{\text{O}}{\overset{\|}{\text{C}}}-$	—	lyotropic nematic	

Table 2.2 Crystallizable Condensation-Type Polymers (continued)

Name	Repeat unit	T_g (°C)	T_m (°C)	T_i (°C)
poly(m-phenylene isophthalamide)		—	390	—
polybenzothiazole		—	lyotropic nematic	
polybenzoxazole		—	lyotropic nematic	
polybenzimidazole		—	lyotropic nematic	
poly(ethylene-p-terphenylate)		—	smectic 322[c]	393[c]
poly(tetraoxyethylene p-terphenylate)		—	smectic 112[d]	249[d]
2-hydroxypropylcellulose [R = CH₂CHCH₃]		—	nematic 110[e]	195[e]
4-hydroxybenzoate/6 hydroxy napthenate copolymer (73/27 mole %)		—	nematic 130[f]	320[f]

[a] All values of T_g, T_m and T_i were taken from Ref. [8], unless noted otherwise.
[b] Ref. [10].
[c] Ref. [11].
[d] Ref. [12].
[e] Ref. [13].
[f] Ref. [14].

do not crystallize. Epoxy resins are an example of this type of system. Also, the presence of a three-dimensional covalent network precludes any further molding, extrusion or drawing.

2.3   Polymers from Ring Compounds

Various small molecules with ring configurations can be polymerized to form crystalline linear chain products. Some polymers obtained in this fashion are listed in Table 2.3. Included in this group are polymers containing silicon and phosphorous atoms in the chain. Polymers formed from ring monomers generally have flexible chains.

Table 2.3   Crystallizable Polymers from Ring Compounds[a]

Name	Repeat unit sequence	T_g (°C)	T_m (°C)
polyoxymethylene	$- CH_2 - O - CH_2 - O -$	-80	183
poly(ethylene oxide)	$- CH_2 - CH_2 - O - CH_2 - CH_2 - O -$	-67	66
isotactic polyepichlorohydrin	$- O - CH_2 - CH - O - CH_2 - CH -$ $\qquad\qquad\ \ CH_2Cl \qquad\qquad CH_2Cl$	—	121
polycaprolactam	$- NH(CH_2)_5 - \overset{\overset{\displaystyle O}{\|}}{C}NH - (CH_2)_5 - \overset{\overset{\displaystyle O}{\|}}{C} -$	—	223
poly(dimethyl silylene)	$-\overset{\overset{\displaystyle CH_3}{\diagup}}{\underset{\underset{\displaystyle CH_3}{\diagdown}}{Si}} -$	—	—
poly(bis-trifluoroethoxy-phosphazene) (R = CH_2CF_3)	$\overset{OR\ \ \ OR}{- N = P - N = P -}\underset{OR\ \ \ OR}{}$	-66	242

[a] All T_g and T_m values were taken from Ref. [8].

2.4   Thermal Properties

The existence of a polymeric system as a rigid glassy liquid, a mobile liquid, a microcrystalline solid or a liquid crystalline mesophase depends on the temperature and the chemical structure of the polymer. Changes from a microcrystalline state to a liquid crystalline or isotropic liquid state takes place at the equilibrium melting temperature.

As seen from the examples given in Tables 2.1, 2.2 and 2.3, T_m and T_g vary widely with a change in the chemical structure. The presence of amide and of aromatic groups in the chain raise T_m and T_g. The morphology of a thermoplastic crystallizable homopolymer at a particular use temperature depends on T_m, which is in turn dependent on the intermolecular forces. If the use temperature is greater than T_m for a crystallizable polymer, only a rubbery liquid morphology will be realized. At temperatures below T_m but above T_g such a material will be partially crystalline, when crystallized quiescently, with rubbery interlayers. Below T_g, the interlayers between crystallites will be glassy.

The semi-rigid aromatic polyamides and the rigid chain polybenzazoles do not form liquid states in the absence of degradation; however, they can be dissolved in highly polar solvents. As noted, many of the aromatic polymers listed in Table 2.2 form thermotropic liquid crystalline mesophases at T_m or lyotropic mesophases in an appropriate solvent. For those forming thermotropic mesophases, the temperature at which the liquid crystalline phase is replaced by an isotropic liquid, T_i, is given, in addition to T_m. A detailed discussion of liquid crystalline mesophases is to be found in Chapter 4.

Two T_g's are given for polytetrafluoroethylene in Table 2.1. At temperatures above the lowest T_g, motion, attributed to a small number of CF_2 groups in the noncrystallized chain sections, is detected. Above the higher of the two T_g's, larger portions undergo conformational changes [3]. Two T_m's are listed for polytetrafluorethylene and for trans-1,4-polybutadiene. Above the lower temperature disorder is present in the crystal structure with the appearance of limited chain motion occurring; an isotropic liquid is observed above the higher temperature given for each substance [4, 15].

Many of the polymers listed in the tables crystallize in more than one atomic array; the melting points refer to the crystal form with the highest T_m. Changes from one form to another at easily attained temperatures and pressures can be reversible or involve melting of one form and crystallization of the other. An example of a reversible change is that for polytetrafluoroethylene at 19 °C from a triclinic form, present below 19 °C, to a trigonal crystal form, present above 19 °C [15]. An example of the second type of transformation is the change of trans-1,4-polyisoprene from an orthorhombic lattice to a monoclinic one [16].

Cooling does not lead to the reverse transformation. These two crystal forms can exist together without spontaneous conversion to only one form.

In Table 2.3, there is no T_m listed for poly(dimethyl silylene); this polymer shows first-order transitions at 160 °C and 220 °C, both of which are classified as intermolecular and not due to crystalline melting [17].

Some polymers with few chain irregularities, although intrinsically crystallizable, can be easily supercooled, without appreciable crystallization, into a glassy amorphous state upon rapid cooling from the melt to a temperature below T_g. Polymers showing this type of behavior usually contain rings in the main or side chains. Examples are poly(ethylene terephthalate), isotactic polystyrene, and various polymers that form liquid crystalline mesophases. These supercooled materials can be crystallized by heating to a temperature where the polymer is below T_m but above T_g. Sufficient time for the various portions of the chains to adopt the conformation necessary for crystallization is then supplied. These types can be crystallized in the presence of a solvent. Extruded poly(ethylene terephthalate) (PET) forms an amorphous film which crystallizes as spherulites when treated with a plasticizing liquid, methylene chloride at 22 °C, a temperature below T_g [18]. Exposure of drawn amorphous PET film to allyl amine liquid leads to crystallization in a fibrous form [19]. Exposure of noncrystallized isotactic polystyrene film to hexahydroindan or cyclooctane vapor at ambient temperatures leads to crystallization [20].

2.5   Lattice Imaging

Crystal lattice images for various polymers have been obtained under high resolution conditions using transmission electron microscopy (TEM) and computerized image enhancement [21]. TEM images taken of small areas can give views of the crystal lattice along the chain direction. The spacings between chains in these lattices, measured from the image, are found to be the same as or somewhat larger than those obtained from X-ray analysis [22, 23]. The images given show these lattices to be defect-free. TEM images giving a view of the lattice parallel to the chain direction, showing the end surfaces, have also been obtained [22, 24].

Scanning tunneling microscopy (STM) has been used to study thin films of both crystallizable and noncrystallizable homopolymers deposited on substrates such as mica and graphite [25–27]. The polymer samples are coated with a thin conducting layer, such as a mixture of platinum and carbon. The same lateral resolution is achieved with STM as with TEM, but STM leads to superior vertical resolution.

Atomic force microscopy (AFM) images have been obtained for a graphite lattice, showing the planar fused-ring structure face-on; these images clearly show the electronic fields of the individual atoms in the crystal lattice [28]. Synthetic polypeptide molecules absorbed on surfaces have been studied with separate chain segments and groups of atoms being clearly apparent [29]. AFM micrographs of oriented chains of various crystallizable homopolymers have been obtained [30–40] . The helical conformation of polytetrafluoroethylene is clearly resolved using AFM [32]. AFM micrographs of a portion of a poly-oxymethylene crystal, with both a raw image and one smoothed by Fourier reconstruction, have appeared [38]. AFM differ-entiates between the O– and CH_2 groups in the polyoxymethylene chain. Two-dimensional Fourier transforms show a larger variation in the chain-to-chain correlation than in the chain periodicity which suggests imperfections in chain packing at the surface of the crystal lattice.

References

1. Bovey, F.A. (1982) *Chain Structure and Conformation of Macromolecules*. New York: Academic Press.
2. Wunderlich, B. (1973) *Macromolecular Physics* Vol. I. New York: Academic Press, p. 388.
3. Woodward, A.E. and Sauer, J.A. (1965) *Physics and Chemistry of the Organic Solid State*, Vol. II. New York: Wiley, p. 691.
4. Ng, S.-B., Stellman, J.M. and Woodward, A.E. (1973) J. Macromol. Sci.-Phys. *7*, 539.
5. Flanagan, R.D. and Rijke, A.M. (1972) J. Polym. Sci. A-2, *10*, 1207.
6. Tanaka, H., Gomez, M.A., Tonelli, A.E., Lovinger, A.J., Davis, D.D. and Thakur, M. (1989) Macromols. *22*, 2427.
7. Miyamoto, Y., Nakafutu, C. and Takemura, T. (1972) Polymer J. *3*, 122.
8. Allcock, H.R. and Lampe, F.W. (1990) *Contemporary Polymer Chemistry*, 2nd Ed. Englewood Cliffs, NJ: Prentice Hall,. pp. 597-607.
9. Wang, P. and Woodward, A.E. (1987) Macromols. *20*, 1818, 1823.
10. Brandrup, J. and Immergut, E.H. (1989) *Polymer Handbook*, 3rd Ed. New York: Wiley.
11. Meurisse, P., Noel, C., Monnerie, L. and Fayolle, B. (1981) Brit. Polym. J. *13*, 55.
12. Fayolle, B., Noel, C. and Billiard, J. (1979) J. Physique *40*, C3-485.
13. Werbowyj, R.S. and Gray, D.G. (1984) Macromols. *17*, 1512.
14. Kaito, A., Kyotani, M. and Nakayama, K. (1990) Macromols. *23*, 1035.
15. Tadokoro, H. (1979) *Structure of Crystalline Polymers*. New York: Wiley.
16. Lovering, E.G. and Wooden, D.C. (1969) J. Polym. Sci. A2 *7*, 1639.
17. Lovinger, A.J., Davis, D.D., Schilling, F.C., Padden, Jr., F.J., Bovey, F.A. and Zeigler, J.M. (1991) Macromols. *24*, 132.
18. Durning, C.J., Rebenfeld, L., Russel, W.B. and Weigmann, H.D. (1986) J. Polym. Sci.: Polym. Phys. *24*, 1341.

19. Qian, R., Shen, J. and Zhu, L. (1981) Makromol. Chem. Rap. Comm. *2*, 499.
20. Tyrer, N.J. and Sundararajan, P.R. (1985) Macromols. *18*, 511.
21. Woodward, A.E. (1988) *Atlas of Polymer Morphology*. Munich: Hanser Publishers.
22. Yeung, P.H.J. and Young, R.J. (1986) Polymer *27*, 202.
23. Dobb, M.G., Johnson, D.J. and Saville, B.P. (1977) J. Polym. Sci.: Polym. Symp. *58*, 237.
24. Chanzy, H.D., Folda, T., Smith, P., Gardner, K.H. and Revol, J.-F. (1986) J. Mat. Sci. Lett. *5*, 1045.
25. Albrecht, T.R., Dovek, M.M., Lang, C.A., Grutter, P., Quate, C.F., Kuan, S.W.J., Frank, C.W. and Pease, R.F.W. (1988) J. Appl. Phys. *64*, 1178.
26. Jandt, K.D., Buhk, M., Petermann, J., Eng, L.M. and Fuchs, H. (1991) Polym. Bull. *27*, 101.
27. Reneker, D.R., Schneir, J., Howell, B. and Harary, H. (1990) Polym. Comm. *31*, 167.
28. Rugar, D. and Hansma, P. (1990) Physics Today, (Oct.), 26.
29. Drake, B., Prater, C.B., Weisenhorn, A.L., Gould, S.A.C., Albrecht, T.R., Quate, C.F., Cannell, D.S., Hansma, H.G. and Hansma, P.K. (1989) Science *243*, 1586.
30. Annis, B.K. and Wunderlich, B. (1993) J. Polym. Sci.: Polym. Phys. *31*, 93.
31. Brinkhus, R.H.G. and Schouten, A.J. (1992) Macromols. *25*, 2717.
32. Dietz, P., Hansma, P.K., Ihn, K.J., Motamedi, F. and Smith, P. (1993) J. Mat. Sci. *28*, 1372.
33. Hansma, H., Motamedi, F., Smith, P., Hansma, P. and Wittmann, J.C. (1992) Polymer *33*, 647.
34. Hanley, S.J., Giasson, J., Revol, J.-F. and Gray, D.G. (1992) Polymer *33*, 4639.
35. Jandt, K.D., Eng, L.M., Petermann, J. and Fuchs, H. (1992) Polymer *33*, 5115.
36. Magonov, S.N., Sheiko, S.S., Deblieck, R.A.C. and Moller, M. (1993) Macromols. *26*, 1380.
37. Snetivy, D. and Vansco, G.J. (1992) Polymer *33*, 432.
38. Snetivy, D. and Vansco, G.J. (1992) Macromols. *25*, 3320.
39. Snetivy, D., Vansco, G.J. and Rutledge, G.C. (1992) Macromols. *25*, 7037.
40. Snetivy, D., Guillet, J. E. and Vansco, G.J. (1993) Polym. Comm. *34*, 429.

3   Morphologies of Crystallized Polymers

The crystallization of flexible chain polymers can be carried out by supercooling a solution or a melt in which, initially, the chain units are either randomly arranged (isotropic fluid) or partially ordered (liquid crystalline mesophase). Crystallization can also occur during the polymerization process of natural or synthetically produced polymers. The crystallization process has to be started or nucleated and can be carried out quiescently or with the application of stress to the fluid. Crystallization from an isotropic polymer fluid is usually incomplete with a sizable portion remaining uncrystallized. The more common morphologies observed include:

1. Faceted single lamellas containing folded or extended chains,
2. nonfaceted lamellas,
3. branched (dendritic) structures,
4. sheaf-like arrays of lamellar ribbons (axialites, hedrites),
5. spherulitic arrays of lamellar ribbons (spherulites),
6. fibrous structures, and
7. epitaxial lamellar overgrowths on microfibrils.

Other morphologies reported include aggregates of curved cup-shaped lamellas and crystallized gels. Further description of these various morphologies will be given below.

3.1   Single Lamellas

Single lamellas (platelets) can be formed by isothermal crystallization from a quiescent supercooled melt or solution of a large variety of polymers with a wide range of chain flexibility [1, 2]. Even polymers with a high degree of chain stiffness, such as cellulose triacetate, are flexible enough to crystallize from solution as single lamellas with folded chains [3].

3.1.1   Geometry

Polymer lamellas can have a highly regular geometry, appearing to be single crystals. Lamellas with uniform size and shape can be prepared by a self-seeding technique involving simultaneous nucleation of all the lamellas and isothermal growth [4]. Faceted lamellar geometries reported include: diamond, hexagonal, pyramidal, square and lathe shapes. Lamellas with rounded edges are also observed. The geometries observed are dependent on the type of crystal lattice adopted, the crystallization solvent and kinetic factors. Some atactic isoregic polymers form irregularly shaped nonfaceted lamellas [5]. Because of their small size, single lamellas grown from solution are usually examined using transmission electron microscopy (TEM), following their deposition on a thin amorphous carbon film and shadowing with a heavy metal, such as platinum or gold, to give contrast.

3.1.2   Chain Folding

In the crystalline part of a lamellar platelet, the polymer chains are usually oriented perpendicular, or nearly perpendicular, to the platelet faces and parallel to the edge surfaces. If the crystal structure is known, the chain orientation can be deduced from selected area electron diffraction on an unshadowed lamella prior to electron microscopy. An electron diffraction photograph from a trans-1,4-polyisoprene lamella is shown in Fig. 3.1. If the molecular weight is low and

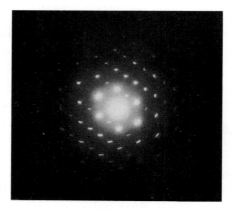

Figure 3.1   Selected area electron diffraction pattern for a singler lamella of trans-1,4-polyisoprene in the β crystal form from solution. Photograph by K. Anadakumaran.

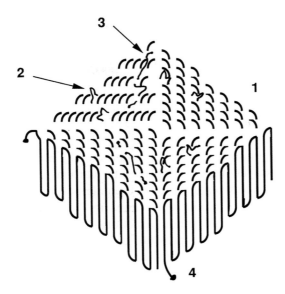

Figure 3.2   Polymer single lamella. The numers identify features described in the text.

the sample essentially monodisperse, completely extended chains packed side-
by-side crystallize from the melt [6]. However, for most synthetic and naturally
occurring polymers the chain length is greater than the lamellar thickness,
requiring the parts of the chain immediately preceding and following a particular
crystal stem to either form folds connecting to other crystal stems or to protrude
from the lamella [7].

A schematic representation of the center part of a polymer lamella, based on
the chain folding model, is shown in Fig. 3.2. In this drawing, each straight line
represents a section of a polymer chain incorporated in the crystal lattice and
contains a number of repeat units; the loops shown depict the folds. Most of the
folds in this figure are tight adjacent re-entrant (region 1). A few of the folds
depicted in Fig. 3.2 are loose adjacent re-entrant (2) and some are nonadjacent
re-entrant (3); some short protruding chain ends (4) are shown.

Both folding and protrusion would lead to a component containing a mixture
of conformations. This noncrystalline component in single lamellas grown from
solution is found by various techniques, such as density, infrared spectroscopy,
Raman spectroscopy and nuclear magnetic resonance, to be appreciable [2]. The
exact size of the noncrystalline component depends on the polymer and the
crystallization conditions. The total number of folds possible per chain depends
on the DP, the crystal stem length, and the fold and chain-end protrusion lengths
[8]. By the use of quantitative chemical reactions, chain folding, possibly

accompanied by short chain-end protrusions, has been shown to definitely occur at lamellar surfaces. Lamellas, containing chains with chemically unsaturated repeat units, such as the polydienes, are reacted in suspension under conditions where only the faces and sides of the lamella undergo change [9–13]. The contribution from the sides to the total number of chain units undergoing reaction is a minor one for the lamellar platelets used. The product that results from reaction of the lamellar surfaces is a block copolymer, containing reacted units (folds and chain-ends) alternating with unreacted units (crystal stems). The average number of monomer units in the unreacted and reacted sequences can be quantitatively assayed using solution state carbon-13 nuclear magnetic resonance. The sequence sizes depend on the polymer and the crystallization conditions. The fraction of chain repeat units in the folds and interlamellar traverses can vary widely; amorphous fractions from near zero for samples crystallized from the melt under pressure to 85% for statistical copolymers containing 10% foreign repeat units have been reported [2, 14]. Deterioration in the morphology, as observed with scanning electron microscopy (SEM) and TEM, occurs with increasing amorphous fraction.

Self-decoration has been used to enhance surface features on single lamellas grown from the melt [15]. Rapid quenching from the crystallization temperature, before all of the polymer molecules in the melt have become involved in lamella formation, leads to the nucleation and growth of small spherulites at lamellar protrusions and edges. Extended chains do not self-decorate to a significant extent, but folded chains do show this effect. Various patterns of undecorated, moderately decorated and heavily decorated regions can arise.

Decoration of polyethylene single lamellas with vaporized tellurium at temperatures greater than 80 °C gave whisker-like growths with the [001] crystal lattice direction oriented along the [020] and [110] crystallographic direction at the fold surface, suggesting the presence of ordered chain folding [16].

Decoration using degraded polyethylene, crystallized as very small lamellar strips from the vapor, has been employed to study the fold surfaces of single lamellas and to investigate crystallization from the melt on substrates at high crystallization temperatures. With polymer lamellas that exhibit growth sectors, the small strips of crystallized evaporated polyethylene are oriented parallel to each other in a direction dependent on the growth surface covered [17]. This has been interpreted as evidence for the occurrence of regular re-entrant chain folds. For lathe-shaped lamellas a more erratic pattern is observed in the middle of the lathe. Polyethylene decoration has been used to study long curved and branched polyethylene lamellas crystallizing from thin films on mica at relatively high temperatures [18]. Curvature generally occurs from one edge of the lamella and ridges up to 15 nm high, identified as edge-on lamellas, are found near the curved edge. The decoration pattern was different for the part of the surface next to a straighter edge than for that part next to the faster-growing curved lateral edges, suggesting the presence of two kinds of fold surfaces.

3.1.3   Dimensions

The thickness of lamellar platelets of polymers is usually small with respect to their lateral dimensions [7]. Thicknesses of 5–20 nm are frequently observed for lamellas from solution and from the melt at 1 atm pressure, while the lateral dimensions are usually in the range of 1–20 μm (1,000–20,000 nm). However, under some special conditions, lamellas visible to the eye can be prepared [19]. Lamellar crystals up to 2 μm in thickness and with areas of up to 1 cm^2 have been prepared during the polymerization of some polydiacetylene materials.

One very important parameter determining the crystal stem length is the degree of supercooling, given by $T_m - T_c$, where T_m is the equilibrium melting point and T_c is the crystallization temperature. The maximum crystalline stem length possible is the length of the extended chain. Crystallization conditions leading to large stem lengths include high pressure, small degrees of supercooling and long crystallization time. The lamellar thickness, a parameter larger than the crystal stem length, is found experimentally to be inversely proportional to the degree of supercooling at low amounts and, then, to level off at high amounts [2]. Melting of polymer lamellas can occur at temperatures well below T_m; the smaller the lamellar thickness, the lower the melting temperature.

An interference contrast optical photomicrograph, showing a group of poly(ethylene oxide) lamellas grown isothermally from the melt at a low degree of supercooling, is given in Fig. 3.3 [6]. The thickness of the middle portion of each lamella corresponds to the length of completely extended chains. Therefore, there are no folds in this center portion. The thickness of the outer parts of the lamellas is about half that of the center part, leading to the conclusion that the chains in the outer regions are folded once. Under the conditions normally used to crystallize synthetic polymers, the crystal stem lengths are many times shorter than the average chain length; thus, multiple folds occur in most chains.

3.1.4   Growth Sectors

Some single lamellas are composed of sectors, triangular-shaped zones with apexes usually meeting at the center of the lamella (see Fig. 3.2) [20]; diamond-

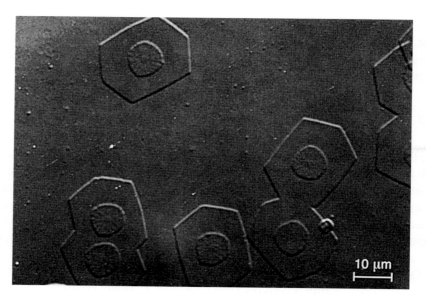

Figure 3.3   Interference contrast optical micrograph of poly(ethylene oxide) lamellas crystallized from a thin film at 58.8 °C. M_w = 6300, M_w/M_n = 1.05. Courtesy of A.J. Kovacs [Reprinted with permission from: A.J. Kovacs and A. Gonthier (1972) "Crystallization and Fusion of Self-seeded Polymers," Kolloid-Z. u. Z. Polymere *250*, 530].

and square-shaped lamellas have four sectors, and hexagonally shaped ones have six. These sectors correspond to different crystallographic growth planes. The addition of more chains to the sectored crystal increases the lamellar dimensions in directions perpendicular to the chain direction. Sectored polymer lamellas can be nonplanar and therefore collapse features, such as creasing, splitting and buckling can be evident in samples deposited from solution [20]. Optical microscopic examination of single lamellas in suspension has confirmed the pyramidal habits taken up under some crystallization conditions. When these samples are prepared for TEM, deposition onto a carbon-coated grid and drying cause collapse of the hollow pyramid [7]. Lathe-shaped lamellas primarily grow along the long dimension of the lathe and sectoring is not observed, although growth by microsectoring has been proposed [17]. Twinning—growth of two lamellas from a common face—can occur; three sets of twins are seen in Fig. 3.3.

3.1.5   Edge and Screw Dislocations

Edge and screw dislocations in a crystal lattice are generated by line translations, as shown in Fig. 3.4. To create an edge dislocation, an extra row of repeat units must be introduced into part of the lattice; to generate a screw dislocation, a slip of part of the lattice occurs. The existence of a screw dislocation in polymer lamellas with a size of about 10 nm is readily observed due to the formation of spiral growths [7]. Crystallization is nucleated at the new side surface in the parent lamella created by the lattice slip. Initially, growth of the new layer occurs away from this line on top of the parent lamella. Other growth planes form and the torsional displacement produces a spiral with the screw dislocation at the center. Either left- or right-handed screw dislocations can form; pairs of these can result in the appearance of a growth terrace [2]. Spirals on a poly(butene-1) lamellar structure grown from solution are seen in the TEM micrograph in Fig. 3.5 [21]. Spiral growths are frequently found when crystallization is carried out nonisothermally or at high degrees of supercooling.

Edge and screw dislocations the size of a unit cell are observed on polymer lamellas using Moiré patterns generated in dark field TEM by two superposed but misaligned lamellas [2]. Dislocation networks have also been viewed using Moiré patterns.

Gold atom decoration of surfaces of single lamellas of homopolymers and of block copolymers grown from dilute solution has been carried out using vaporized metal [4, 22]. Growth steps on the lamellas are observed; a clustering of the heavy metal on portions of some of the lamellar surface is seen, particularly after the decorated surface is washed with crystallization solvent.

Lamellar overgrowths, smaller in size than the parent lamella and connected thereto, and apparently nucleated by protruding chains, rather than by screw

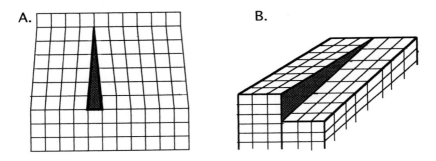

Figure 3.4   Translational defects: A. edge dislocation, B. screw dislocation.

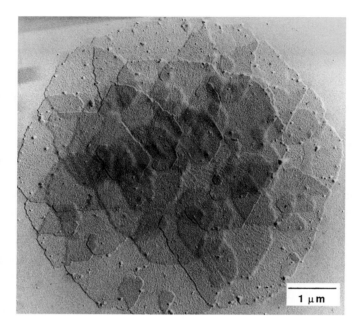

Figure 3.5   Transmission electron micrograph showing growth spirals on a poly-butene-1 lamella crystallized from 0.01% pentyl acetate solution at 50 °C. [Reprinted with permission from: A.E. Woodward and D.R. Morrow, (1968) J. Polm. Sci. A-2 6, 1987. Copyright © 1968, John Wiley and Sons, Inc.]

dislocations, have been observed for various polymers. Lamellas containing such overgrowths are distinct from, but are possibly related to, the multilamellar structures discussed in the next section.

3.1.6   Thermally Induced Changes

When isolated polymer crystal lamellas are annealed at temperatures above the crystallization temperature, but below the equilibrium melting temperature, the lamellar thickness increases and holes appear, as observed using TEM [2, 23]. One example of lamellar thickening due to annealing is shown in Fig. 3.6 for isotactic polypropylene [24]. Original crystal outlines are evident in the figure, and measurements of shadow lengths show that the remaining material has a thickness greater than that for the original lamellas.

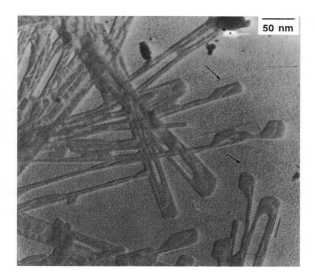

Figure 3.6   Transmission electron micrograph of isotactic polypropylene lamellas crystallized from solution, dried and heated to 160 °C for 1 hr (arrows point to original crystaloutlines). Pt-shadowed. Courtesy of G.C. Richardson.

Transitions other than melting (and recrystallization) can occur upon heating single lamellas. A thermochromic phase transition (blue to red) occurs at about 115 °C for large single crystals of the polydiacetylene obtained by solid state polymerization of 5,7-dodecadiyne-1,12-diol bis(ethyl urethane) [25]. After melting and recrystallization, spherulites, which are multilamellar structures, appear; changes in the crystal structure, due to main- and side-chain conformational changes, are observed [26].

3.2   Branched and Multilamellar Structures

The formation of single lamellas during quiescent crystallization of a polymer with flexible chains from solution or the melt, as discussed above, occurs in prenucleated systems at relatively low degrees of supercooling. Under many conditions, particularly during crystallization from concentrated solution or the melt, branched and/or multilayered lamellar structures appear. These structures which include dendrites, hedrites (or axialites) and spherulites will be discussed separately below.

3.2.1   Dendrites

Dendrites are branched lamellas with the branches related in a definite crystallographic way to each other. Branching occurs under conditions in which crystal growth from solution is fast relative to the diffusion of polymer molecules to the crystallization site. This results in a lowering of the polymer concentration at the growth faces. Dendritic structures are found for solution crystallized polyethylene, with branches along the crystallographic a and b axes, perpendicular to the chain direction in the lamellas [27]. A schematic representation is given in Fig. 3.7A. These structures can have lengths of hundreds of microns [28]. At high concentrations, thick multilayered dendrites are observed. Interpenetrating cross-hatched dendritic structures, called quadrites, of one polymer, isotactic polypropylene, have been crystallized from solution and from the melt [29, 30]; a drawing of part of one of these is shown in Fig. 3.7B.

3.2.2   Hedrites

Hedrites (axialites) are stacks of centrally connected lamellas that grow from solution or the melt [1]. The lamellas in a particular stack are probably commonly nucleated by one or more chains with elongated conformations. (This form of nucleation will be discussed in more detail below). The lamellas are usually somewhat longer than they are wide, with growth occurring principally at the ends. During growth from solution, these lamellas can develop further connections due to the incorporation of a single chain or a group of chains in neighboring lamellas. One consequence is that two lamellas can be closely connected to each other at more than one place in the stack; repeated events of this sort give the hedrite a cellular appearance. A lamellar stack can exhibit considerable curvature, mainly to one side but along the width and length of the lamellas, causing the stack to appear cup-shaped. Separation or splaying of the lamellar ribbons can occur. The details of hedrites from solution are seen using SEM [23, 31]; these features can be preserved and enhanced by reaction with a heavy metal containing compound, such as OsO_4 or RuO_4, before solvent re moval. When crystallization is carried out at high solution concentrations, gels containing hedrite-like structures can be obtained [32, 33]. Crystallization from a supercritical fluid yields hedrites [34]. Branching, cellulation and splaying are observed in these structures, as shown by SEM in Fig. 3.8.

Hedrites are usually birefringent when viewed through crossed polaroids. When viewed face-on in suspension, they show an oval of light, the central portion being dark due to insufficient thickness. When viewed from the edges of

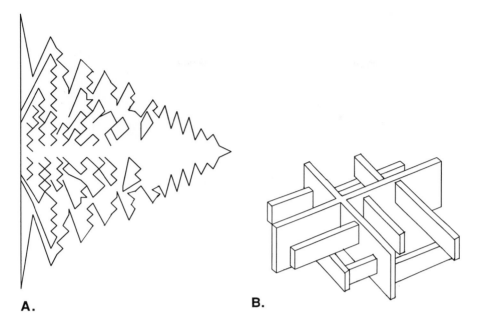

A.   **B.**

Figure 3.7   Polymer dendrites (partial views): A. polyethylene, B. polypropylene.

Figure 3.8   Scanning electron micrographs of polyethylene hedrites grown from supercritical propane solution. Courtesy of P. Ehrlich. [Reprinted with permission from: Bush, P.J., Pradhan, D., and Ehrlich, P. (1991) Macromols. *24*, 1439. Copyright © 1991, American Chemical Society].

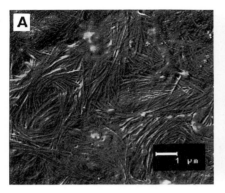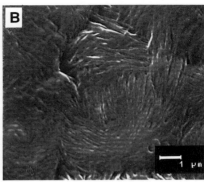

Figure 3.9   Scanning electron micrograph of melt crystallized trans-1,4-polyisoprene hedrites: A. $T_c = 0$ °C, B. $T_c = 36$ °C. Photographs by N. Vasanthan.

the stack in suspension, four thick, bright quadrants are usually seen and, in some cases, six bright parts are observed [8]. Rotation of the stack during observation brings about rotation of the bright (and dark) portions of the pattern.

The crystal stems in a hedrite are perpendicular, or nearly so, to the faces of the ribbons; these faces contain chain folds. The amorphous fraction in hedrites depends on the molecular weight and at high molecular weights is 20% larger than that in single lamellas [13]; this larger amorphous fraction is believed to be due mainly to exposed sections of chains connecting the crystal stems in two adjacent lamellas [35].

Hedrites can also be observed upon crystallization from the melt [2]. SEM micrographs of melt-crystallized hedrites of trans-1,4-polyisoprene (TPI) are shown in Fig. 3.9 [36]. These hedrites in the β crystal form, were prepared using unfrac-tionated polymer crystallized from thin films by: 1) quenching to 0 °C and heating to 25 °C and 2) isothermal crystallization at 36 °C followed by cooling to 25 °C. Treatment with OsO_4 was carried out prior to SEM. The lamellas in these melt-crystallized structures are more tightly packed than those in hedrites grown from solution where gaps of 1–5 μm can occur [23]. Considerable lamellar curvature is observed in melt-crystallized TPI hedrites. The hedrites grown from the melt impinge on one another and are joined together by lamellas, although some spaces appear between hedrites, at least near the surface of the sample film. The number of lamellas coupled in each multilamellar unit decreases with increasing degree of supercooling (lowering of the crystallization temperature), as can be seen by a comparison of Fig. 3.9A and 3.9B. When melt-crystallized hedrites are investigated with TEM, it is usually necessary to study

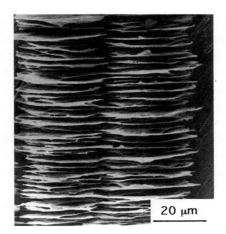

Figure 3.10   Scanning electron micrograph showing lamellar platelets of an aliphatic polyester crystallized in hexagonal channels of a urea lattice. Courtesy of F. Brisse [Reprinted with permission from: Chenite, A. and Brisse, F. (1993) Macromols. 26, 3055. Copyright © 1993, American Chemical Society].

fracture surfaces or thin microtomed surfaces which are etched and then replicated [37]. Hedrites containing over 100 lamellas have been obtained [38].

Crystallization from solution or the melt of stiff chains can yield narrow fibrous lamellas, lamellar bundles or faceted lamellas [39, 40].

3.2.3   Urea/Polymer Adducts

Urea forms hydrogen bonded structures containing hexagonal channels. Polymerization in these channels yields stacks of hexagonal platelets, as is evident from the SEM micrograph shown in Fig. 3.10 [41].

3.2.4   Spherulites

3.2.4.1   Structure and Occurrence

Spherulites are composed of lamellar ribbons that grow outward, radially: 1) from a single (heterogeneous) nucleus, 2) from a group of lamellas similar to that in a hedrite [1, 2, 38], or 3) from a quadrite, a crosshatched lamellar array [29]. After sufficient supercooling of a relatively concentrated solution or of a melt, spherulites can be grown at constant or slowly decreasing temperature. The growing spherulite is filled in by lamellar branching, particularly at high degrees of supercooling. Branches can be created by nucleation of new lamellas at the surface of a parent lamella, as occurs in hedrites, or by dendritic branching, branching at the growth face which occurs at high degrees of supercooling. When the crystallization temperature is raised, the degree of supercooling is decreased, branching diminishes, the primary lamellas increase in average thickness, and additional lamellas fill in the spherulite structures at a later time. Both curvature in and twisting of the lamellas in spherulites, occurring during growth, has been reported [42, 43]. Spherulitic crystallization from the melt usually yields a coherent sample made up of individual structures impinging on and linked to one another by lamellas common to two spherulites. Spherical symmetry is lost when two or more growing spherulites impinge on one another or when crystallization of two different crystal forms occur simultaneously [44].

Although optical microscopic investigation is carried out at low magnifications and gives a two-dimensional view with the spheres appearing as circles, the evolution of the spherulite with time can be followed by this method. In many optical microscopic studies of spherulites, crossed polaroids are used. Due to birefringence in many of these structures, some of the plane polarized light is transmitted through an analyser polaroid, with the same pattern being observed for each spherulite. A common pattern is four bright quadrants separated by a dark cross, with the arms of the cross in the direction of the polarizer and analyser. In addition, sometimes alternating bright and dark rings are observed. Various other birefringence patterns have been reported with spherulites of a particular polymer changing appearance as the crystallization temperature is changed [30, 45–51].

SEM supplies a three-dimensional view of spherulites after growth from the melt, as is shown in Fig. 3.11A for isotactic polypropylene [52]. At the higher magnifications available using SEM, considerable detail concerning the lamellar ends at the spherulite surface can be revealed (Fig. 3.11B).

TEM studies of spherulites of a wide variety of flexible chain polymers have been reported [1, 2, 23]. However, unless thin samples are used, extensive preparation is usually required prior to TEM study. Typically, microtoming or fracture at low temperatures, etching with a destructive reagent to remove amorphous and low molecular weight portions and replication are carried out. Care must be taken during etching to prevent the introduction of artifacts, features that are not present in the original preparation [53], and to properly interpret the results of random slices through spherulites containing radially oriented and twisted lamellas [43]. Reaction of similar polyethylene samples with chlorosulfonic acid and permanganic acid showed that the lamellar overgrowths under study were smaller when treated with chlorosulfonic acid

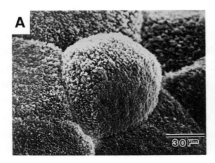

Figure 3.11   Scanning electron micrographs of a fracture surface (liquid N_2) of iso-tactic polypropylene crystallized isothermally from the melt in a glass tube at 110 °C for 5 hr after cooling from 220 °C: A. spherulites, B. spherulite surface. Courtesy M. Kojima [Reprinted with permission from: Kojima, M. (1979) J. Polym. Sci.: Polym. Lett. *17*, 609. Copyright © 1979, John Wiley and Sons, Inc.].

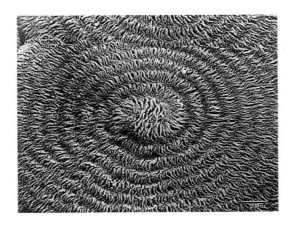

Figure 3.12   Scanning electron micrograph of the center of a polyethylene spherulite ion etched at 500 V(ac), 3 ma for 1 hr in air. Courtesy of J.H. Magill. [Shankernarayanan, M.J, Sun, D.C., Kojima, M., and Magill, J.H. (1987) J. Int. Processing Soc. *1*, 66.]

[54]. The general appearance of trans-1,4-polyisoprene hedrite and spherulite structures from solution viewed in suspension before and after OsO_4 treatment was the same [8]. Ion etching in air or Ar can be used in a similar way to etching with reactive chemicals to preferentially strip away parts of a surface [55–57]. The lamellar nature of polyethylene spherulites are clearly revealed by ion etching at 500 V(ac), 3 ma for 1 hour in air, as seen in Fig. 3.12 [55].

Crystallization at high pressure (4–5 kilobars) and increased temperature yields spherulites with lamellas of greater thickness and length than those obtained at 1 atm. The lamellar thicknesses, measured from TEM micrographs of replicated fracture surfaces are about 2–6 μm for polyethylene spherulites [2]. The chains in these lamellas are highly, if not completely, extended. Part of an extended chain polyethylene spherulite is shown in the TEM micrograph of a replica given in Fig. 3.13 [58]. Kinks in crystal stems have been observed with TEM on fracture surfaces of extended chain lamellas prepared under pressure [2]. Kinks can also be introduced into single lamellas by stress application, as discussed in Chapter 7.

Crystallization from solution can yield separate highly branched structures with a lamellar stack at the center [2, 23]. These structures can be relatively open while suspended in the crystallization liquid. When the liquid is removed, collapse takes place. Aggregates of interconnected cup-shaped lamellas and relatively tightly packed structures of twisted and interconnected lamellas have been observed [8].

Multilamellar structures from solution are large enough to be viewed using optical microscopy while they are in suspension. The spherically symmetrical structures usually show irregular light and dark areas when viewed using crossed polaroids. More detailed micrographs can be obtained using SEM. However, to avoid structural collapse upon drying, it is best to freeze or fix the structure by reaction with a nondestructive chemical; this treatment may also improve the contrast. If the structure is thin enough, TEM can be employed to investigate morphological details following heavy metal shadowing.

When polymer samples are crystallized at elevated temperatures and then cooled to 25 °C, the amount of crystallinity increases during the cooling process. This could either involve the nucleation of additional lamellar structures or be due to an increase in order of the amorphous portions of chains associated with crystal stems. The presence of some additional, smaller lamellar structures after cooling from T_c to 25 °C have been observed using SEM for a crystallizing system where the melting temperature is close to 25 °C [59]. However, infrared spectroscopy evidence for the same preparations has been given that supports the concept of chain ordering occurring during the cooling process [60].

Three crystallization regimes, given in terms of the spherulitic growth rate, have been identified [61, 62]. Regime I is found at low degrees of supercooling (high crystallization temperatures) where nucleation events are rare and crystal growth is the main process. At lower crystallization temperatures (higher degrees of supercooling), there is a transition to Regime II in which nucleation is competitive with crystal growth. At still higher degrees of supercooling, Regime III is reached, where nucleation occurs at a greater frequency than

Figure 3.13   Transmission electron micrograph of a fracture surface replica of an extruded chain polyethylene spherulite crystallized at 215–225 °C and 4.8 kb for 10–20 hr and cooled to room temperature under pressure. Courtesy of B. Wunderlich [Reprinted with permission from: Prime, R.B., Wunderlich, B., and Mellilo, L. (1969) J. Polym. Sci. A-2 7, 2091. Copyright © 1969, J. Wiley and Sons, Inc.].

crystal growth and is the dominant process. Spherulites formed in Regime II can show a considerable twisting of the lamellas and the presence of numerous screw dislocations [63].

The specific morphology obtained during polymer crystallization depends on the temperature or the thermal history, if more than one temperature is employed. Spherulites of isotactic polypropylene in the α-crystal form grow from hedrites at 160 °C and from quadrites at 150 °C and below [30]. Polypropylene spherulites with separate shells of the α and γ crystal forms have been grown by changing T_c during crystallization, with the different shells distinguishable under the polarizing optical microscope due to a difference in birefringence [50]. A change from an axialite morphology to a ringed spherulitic one occurs for polyethylene when the crystallization temperature is changed from 128 °C to 121 °C [46]. The spherulitic morphology and the crystal form of poly[(bis-4,4'-dicyclohexyl-methane) n-docecanediamide] crystallized from m-cresol solution under isothermal solvent removal at temperatures from 160 °C to 200 °C depends on T_c [49]; six different spherulitic morphologies and three crystal forms were reported. A reversible change from radiating to banded spherulites occurs at elevated temperature for poly(aryl ether ether ketone) with the radiating form being the stable one at room temperature [51]. Polyimides containing both rigid and flexible groups can form different types of spherulites and in some cases axialites, depending on the length of the flexible group and the crystallization temperature [48].

Spherulite structures with characteristics different from those described above have been reported. Quasi-circular spherulites of one polymer, poly-(vinylidene fluoride) in the a crystal phase, have been seen using thin samples [64]; spherulites of this polymer in the γ crystal form made up of scrolled lamellas have been reported [65].

3.2.4.2   Transcrystallization

Transcrystallization is a phenomenon occurring during spherulitic crystallization in a supercooled melt containing certain fibers [66–71]. Cylindrical columns of crystallized material, with diameters the same as those of the free spherulites present, form around the fiber. Due to crowding on the fiber caused by the nucleation of many potential spherulitic structures, development in the long direction of the fiber is impeded while growth radially occurs in a normal fashion. Whether or not transcrystallization occurs depends on the type of fibrous material present and the crystallization temperature [68, 69].

3.2.4.3 Noncrystallized Fraction

Spherulitic structures grown from the melt usually contain a moderate to large noncrystallized fraction. This fraction is significantly larger than that existing at the surfaces of single solution grown lamellas due to the larger degree of supercooling in melt crystallization and the accompanying smaller crystal stem length. In multilamellar structures not all of the noncrystalline component resides in chain folds and chain ends, particularly when the nucleation rate is large. Under that condition a given chain can be incorporated in more than one lamella, leading to the presence of an interlamellar traverse. For each interlamellar traverse gained, at least one fold is lost. The presence of interlamellar traverses (or tie-chains as they are sometimes called) has been demonstrated for polyethylene [35]. A rigid amorphous interfacial region a few nanometers in thickness, between the crystallites and the isotropic amorphous part, which presumably contains any folds present, is believed to exist in melt-crystallized spherulites [72, 73].

3.2.4.4 Molecular Weight Dependence

Unfractionated samples of crystallizable synthetic homopolymers are generally polydisperse. Hedrites or spherulites usually appear upon melt crystallization of these homopolymers.

TEM studies on fractionated polyethylene and poly(ethylene oxide) show the morphology of isothermally crystallized samples to be dependent on the molecular weight as well as the degree of supercooling [42, 74]. Isothermal crystallization gives axialites at low molecular weights, sheet-like structures at moderate molecular weights, spherulites at moderate to high molecular weights and randomly positioned lamellas at very high ($>2 \times 10^6$) molecular weights.

For a low molecular weight (5.6×10^3) polyethylene fraction crystallized at high T_c, long, roof-shaped lamellas appear with a maximum thickness near that for fully extended chains; lowering T_c causes the lamellar thickness observed for this fraction to markedly decrease [75]. For fractions of moderate molecular weight (7.7×10^4–2.5×10^5) crystallized at high T_c, maximum thicknesses of about 100 nm were found, while for high molecular weight fractions (1.6×10^6–6×10^6) the maximum was about 50 nm and curved as well as roof-shaped crystals occur. When the high molecular weight fractions are crystallized at intermediate T_c, short curved lamellas appear. The crystalline fraction realized decreases with increasing molecular weight [74]. The morphological changes as well as the decrease in lamellar thickness and crystalline fraction with increasing molecular weight are believed to be due to increased chain entanglements.

3.3   Fibrous Structures

Fibers are long, narrow, ribbon-like structures, having a different chain orientation than that present in lamellas. In these structures, the chain (c-axis) orientation is mainly along the ribbon direction; it is not perpendicular to it, as found in lamellas [2, 76]. Fibers are usually composed of smaller units, fibrils, microfibrils and nanofibrils, oriented in the same direction [77]. Fibrillar structures can be formed naturally, during synthesis, from solution, from the melt, from a gel or from quiescently melt-crystallized material, with the chains being placed in extended conformations during the crystallization or recrystallization process. Fibers can be prepared from crystallizable polymers with flexible chains (see Chapter 2 for various examples), as well as from those with completely stiff or semi-stiff chains, such as polybenzothiazole and polybenzamide, respectively.

Depending on the method of preparation, fibers from flexible chain polymers can contain extended noncrystalline portions of chains and chain folds, or they can have perfect or near-perfect order, as shown schematically in Fig. 3.14. A change from a spherulitic morphology to a fibrous morphology can be brought about by the application of a tensile stress above T_g for the polymer, a process called cold drawing; fibrous samples prepared in this way contain chain folds and noncrystalline extended chain portions, as in Fig. 3.14A [78].

A.   B.

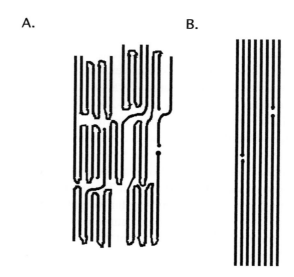

Figure 3.14   Fibrillar crystallization: A. cold-drawn melt crystallized, B. ultradrawn solution crystallized.

Crystallization of a flexible chain polymer from solution can produce a fibrous precipitate if a high rate of stirring is applied and the crystallization temperature is kept above those used to prepare lamellar crystals in quiescent solutions. The fibrils formed in this manner are not generally aligned with respect to one another [23].

Perfect or near-perfect order can be obtained by drawing solution crystallized samples to high degrees [77, 79]. Imperfections in superdrawn fibers of polyoxymethylene, a flexible chain polymer, have been reported [80]; this material is composed of a network of thick fibrils connected by thin cross fibrils creating chains of voids.

Highly extended and aligned fibers of semi-rigid (aromatic polyesters and polyamides) are prepared by crystallization from ordered liquid crystalline solutions [79, 81]. Rigid chain polymers, such as polybenzimidazole and polybenzoxazole, are completely extended and form fibers from solution.

Crystallization from a nematic mesophase of semi-stiff chains of a poly(4-hydroxybenzoate)/poly(6-hydroxy-4-napthenate) copolymer can yield lamellar stacks linked together by extended chains [82]. The lamellas are not apparent with TEM unless etching precedes replication. Using a single step replication technique, oriented lamellar structures are observed. It is believed that the chains connecting adjacent crystallites, as crystallized, undergo axial shifts.

Crystallization of the stiff backbone polyimide with the following structure:

has been accomplished by casting from solution [40]. However, if the solvent is evaporated in a moist atmosphere prior to onset of crystallization, amorphous beads of material appear in place of the crystallized product.

The degree of orientation in a fiber can be determined by birefringence measurements using an optical microscope or by X-ray analysis. Using TEM and SEM, the microfibril diameter is found to be about 50 nm with an orientation that is parallel to the fiber direction [83]. The absence of long-range lateral order in a cellulose fibril was shown using atomic force microscopy (AFM) [84]. Investigation of a gel-drawn ultrahigh molecular weight polyethylene identified the elementary building block in a fiber, the nanofibril, and showed that the microfibril diameter (0.2–1.2 μm) and the bundle size (4–7 μm) decreased with increasing tensile elongation [77].

3.4   Epitaxial Crystallization

Epitaxial or oriented growth of flexible chain polymer crystals on crystalline substrates, such as mica, NaCl or KBr, can occur from solution or the melt. The substrate is cut to expose a particular crystal plane prior to polymer crystal growth. Needle-shaped growths containing folded chains and with growths oriented in certain directions, as directed by the crystallography of the substrate, are observed. In some cases, such as isotactic polypropylene on mica, the structures are highly branched dendrites [85].

Epitaxial crystallization of a polymer on the surface of a highly oriented crystalline film of another polymer has been carried out by pressing thin films of the two polymers together and applying a tensile stress during the melt crystallization process [86–89]. Using thin films of polypropylene and polyethylene and annealing the product gives a cross-hatched structure of polyethylene lamellas oriented at 45° to the orientation direction in the polypropylene [86].

The nucleation of folded chain lamellar growth on a small group of extended chains of the same or a different polymer is another form of epitaxial growth [2]. In this type of crystallization the c crystallographic axis, usually the chain axis, in the lamellas and in the fiber are parallel. A schematic representation of folded chain growth on an extended chain is shown in Fig. 3.15. Epitaxial crystallization can take place in stirred solutions [90]; the lamellas, nucleated on the extended chains, are separated and easily distinguished. (Due to their appearance these structures are referred to as shish-kebabs.) Lamellar crystallization of one

Lamellas

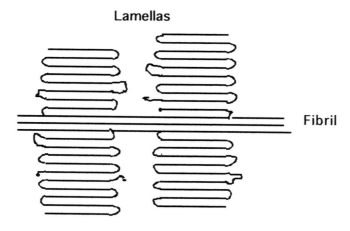

Fibril

Figure 3.15   Epitaxial lamellar crystallization on polymer fibrils.

Figure 3.16 Transmission electron micrograph of (fractionated) mannan lamellas epitaxially grown on cellulose fibrils from dilute NaOH solution at elevated temperature by addition of dimethylsulfoxide. Courtesy of H. Chanzy and M. Dube [Reprinted with permission from: Chanzy, H., Dube, M., Marchessault, R.H., and Revol, J.F. (1979) Biopolymers *18*, 887. Copy-right © John Wiley and Sons, Inc.].

polymer on fibers of another can occur, as shown in Fig. 3.16 [91]; the epitaxial growth of the natural polymer, mannan, on cellulose fibrils during crystallization from solution is shown in this figure.

As pointed out above, crystallization from quiescent solution can produce groups of closely spaced lamellas, called hedrites; these are believed to be nucleated in the same way as the shish-kebab structures. The nucleation of folded chain lamellar growth on extended chains can occur in crystallization from the melt, particularly upon application of a uniaxial tensile stress [92], a shear stress [50] or during extrusion. In these cases the lamellas are close together or inter-locked, respectively.

Under certain conditions, epitaxial growth of folded chain lamellas would be considered a polymer decoration effect. As an example, the chains at the surfaces of exposed extended chain lamellas have been shown to support epitaxial growth of folded chain lamellas [58]. The chain direction is disclosed due to this decoration.

3.5   Other Morphologies

Crystalline polymer morphologies also include two-dimensional matrices, as found for graphite and for mica, and regular three-dimensional networks, present in diamond.

References

1. Geil, P.H. (1963) *Polymer Single Crystals*. New York: Wiley.
2. Wunderlich, B. (1973) *Macromolecular Physics, Vol. 1*. New York: Academic Press.
3. Chanzy, H.D., Taylor, K.J., and Vuong, R. (1988) in *Atlas of Polymer Morphology*. Munich: Hanser Publishers, pp. 25, 47.
4. Blundell, D.J. and Keller, A. (1973) J. Macromol. Sci.: Phys. *B7*, 253.
5. Lovinger, A.J. and Cais, R.E. (1984) Macromols. *17*, 1939.
6. Kovacs, A.J. and Gonthier, A. (1972) Kolloid-Z. u. Z. Polymere *250*, 530.
7. Keller, A. (1958) in *Growth and Perfection of Crystals, Proc. Int. Conf. Crystal Growth, Cooperstown*. New York: Wiley, pp. 499-528.
8. Kuo, C.-C. and Woodward, A.E. (1984) Macromols. *17*, 1034.
9. Stellman, J.M. and Woodward, A.E. (1969) J. Polym. Sci. *7*, 755.
10. Schilling, F.C., Bovey, F.A., Tseng, S., and Woodward, A.E. (1983) Macromols. *16*, 808.
11. Schilling, F.C., Bovey, F.A., Anandakumaran, K., and Woodward, A.E. (1985) Macromols. *18*, 2688.
12. Wang, P-G. and Woodward, A.E. (1987) Macromols. *20*, 1818.
13. Xu, J. and Woodward, A.E. (1988) Macromols. *21*, 83.
14. Vasanthan, N., Corrigan, J.P., and Woodward, A.E. (1994) Makromol. Chem. (in press).
15. Kovacs, A.J., Gonthier, A., and Straupe, C.(1975) J. Polym. Sci. *C50*, 283.
16. Faghihi, S., Hoffman, T., Petermann, J. and Martinez-Salazar, J. (1992), Macromols. 25, 2509.
17. Wittmann, J.C. and Lotz, B. (1985), J. Polym. Sci.: Polym. Phys. Ed. 23, 205.
18. Keith,H.D., Padden Jr., F.J., Lotz, B., and Wittmann, J.C. (1989) Macromols. *22*, 2230.
19. Thakur, M. and Meyler, S. (1985) Macromols. *18*, 2341.
20. Bassett, D.C., Frank, F.C., and Keller, A. (1963) Phil. Mag. *8*, 1753.
21. Woodward, A.E. and Morrow, D.R. (1968) J. Polym. Sci. A2 *6*, 1987.
22. Kojima, M. and Magill, J.H. (1974) J. Macromol. Sci.: Phys. *B10*, 419.
23. Woodward, A.E. (1988) *Atlas of Polymer Morphology*. Munich: Hanser Publishers.
24. Richardson, G.C. (1988) in *Atlas of Polymer Morphology*. Munich: Hanser Publishers, p. 161.
25. Tanaka, H., Thakur, M., Gomez, M.A., and Tonelli, A.E. (1987) Macromols. *20*, 3094.
26. Tanaka, H., Gomez, M.A., Tonelli, A.E., Lovinger, A.J., Davis, D.D., and Thakur, M. (1989) Macromols. *22*, 2427.
27. Fischer, E.W. and Lorenz, R. (1963) Kolloid Z. u. Z. Polym. *189*, 97.
28. Wunderlich, B. and Sullivan, P. (1962) J. Polym. Sci. *61*, 195.
29. Sauer, J.A., Morrow, D.R., and Richardson, G. C. (1965) J. Appl. Phys. *36*, 3017.
30. Olley, R.H. and Bassett, D.C. (1989) Polymer *30*, 399.
31. Xu, J. and Woodward, A.E. 1 (1986), Macromols. 19, 1114.

32. Mandelkern, L., Edwards, C.O., Domszy, R.C., and Davidson, M.W. (1985) *Microdomains in Polymer Solutions.* New York: Plenum Press, pp. 121–141.
33. Wang, P-G. and Woodward, A.E. (1987) Macromols. *20*, 2718.
34. Bush, P.J., Pradhan, D., and Ehrlich, P. (1991) Macromols. *24*, 1439.
35. Keith, H.D., Padden Jr, F.J., and Vadimsky, R.G. (1966) J. Polym. Sci. A-2 *4*, 267.
36. Vasanthan, N. (1993) Unpublished results.
37. Bassett, D.C. and Vaughan, A.S. (1985), Polymer 26, 717.
38. Rensch, G.J., Phillips, P.J., Vatansever, N., and Gonzalez, A. (1986) J. Polym. Sci.: Polym Phys. *24*, 1943.
39. Lovinger, A.J., Hudson, S.D., and Davis, D.D. (1992) Macromols. *25*, 1752.
40. Waddon, A.J. and Karasz, F.E. (1992) Polymer *33*, 3783.
41. Chenite, A. and Brisse, F. (1993) Macromols. *26*, 3055; also (1992) Macromols. *25*, 776.
42. Bassett, D.C. and Hodge, A.M. (1981) Proc. R. Soc. London A *377*, 39.
43. Lustiger, A., Lotz, B., and Duff, T.S. (1989) J. Polym. Sci.: Polym. Phys. *27*, 561.
44. Schulze, G.E.W. and Wiberg, H.-P. (1989), Colloid and Polym. Sci. 267, 116.
45. Cheng, S.Z.D., Barley, J.S., and von Meerwall, E.D. (1991) J. Polym. Sci.: Polym. Phys. *29*, 515.
46. Chew, S., Griffiths, J.R., and Stachurski, Z.H. (1989) Polymer *30*, 874.
47. Chiu, G., Alamo, R.G., and Mandelkern, L. (1990) J. Polym. Sci.: Polym. Phys. *28*, 1207.
48. Heberer, D.P., Cheng, S.Z.D., Barley, J.S., Lien, S.H.-S., Bryant, R.G. and Harris, F.W. (1991), Macromols. 24, 1890.
49. Li, L.S. and Geil, P.H. (1991) Polymer *32*, 374.
50. Varga, J. (1992) J. Mat. Sci. *27*, 2557.
51. Zhang, Z. and Zeng, H. (1992), Makromol. Chem. 193, 1745.
52. Kojima, M. (1979) J. Polym. Sci.: Polym. Lett. *17*, 609.
53. Naylor, K.L. and Phillips, P.J. (1983) J. Polym. Sci.: Polym. Phys. *21*, 2011.
54. Bashir, Z., Hill, M.J., and Keller, A. (1986) J. Mat. Sci. Lett. *5*, 876.
55. Shankernarayanan, M.J., Sun, D.C., Kojima, M., and Magill, J.H. (1987) J. Int. Polym. Processing Soc. *1*, 66.
56. Kojima, M. (1988) in *Atlas of Polymer Morphology.* Munich: Hanser Publishers, pp. 474, 519.
57. Chen, S.S., Hu, S.R., and Xu, M. (1988) in *Atlas of Polymer Morphology*, Munich: Hanser Publishers, pp.474, 521.
58. Prime, R.B., Wunderlich, B., and Melillo, L. (1969) J. Polym. Sci. A2 *7*, 2091.
59. Woodward, A.E., Corrigan, J.P., and Vasanthan, N. (1993) Trends in Polymer Science *3*, 299.
60. Vasanthan, N., Corrigan, J.P., and Woodward, A.E. (1993) Polymer *34*, 2270 (1993).
61. Hoffman, J.D. (1983) Polymer *24*, 3.
62. Lovinger, A.J., Davis, D.D., and Padden, Jr., F.J. (1985) Polymer *26*, 1595.
63. Phillips, P.J. and Philpot, R.J. (1988), in Atlas of Polymer Morphology, Hanser Publishers, Munich, pp. 93, 119.
64. Lovinger, A.J. (1980) J. Polym. Sci.: Polym. Phys. *18*, 793.
65. Vaughan, A.S. (1993) J. Mat. Sci. *28*, 1805.
66. Avella, M., Della Volpe, G., Martuscelli, E., and Raimo, M. (1992) Polym. Eng. and Sci. *32*, 376.
67. Chatterjee, A.M., Price, F.P., and Newman, S. (1975) J. Polym. Sci.: Poly. Phys. Ed. *13*, 2369.
68. Ishida, H. and Bussi, P. (1991) Macromols. *24*, 3569.

69. Thomeson, J.L. and van Rooyen, A.A. (1992) J. Mat. Sci. *27*, 889.
70. Wang, W., Qi, Z. and Jeronimidis, G. (1991) J. Mat. Sci. *26*, 5915.
71. Yue, C.Y. and Cheung, W.L. (1991), J. Mat. Sci. 26, 870.
72. Tanabe, Y., Strobl, G.R., and Fischer, E.W. (1986) Polymer *27*, 1147.
73. Bader, H.G., Schnell, H.F.E., Goritz, D., Heinrich, U.-R., and Schultze, H.-J. (1992) J. Mat. Sci. *27*, 4726.
74. Mandelkern, L. (1985) Polym. J. *17*, 337.
75. Voigt-Martin, I.G. and Mandelkern, L. (1984) J. Polym. Sci.: Polym. Phys. *22*, 1901.
76. Sawyer, L.C. and Grubb, D.T. (1987) *Polymer Microscopy*. London: Chapman and Hall.
77. Magonov, S.N., Shieko, S.S., Deblieck, R.A.C., and Moller, M. (1993) Macromols. *26*, 1380.
78. Peterlin, A. (1965), J. Polym. Sci. *9C*, 61.
79. Lemstra, P.J., van Aerle, N.A.J.M., and Bashaansen, C.W.M. (1987) Polymer J. *19*, 85.
80. Komatsu, T. (1993) J. Mat. Sci. *28*, 3035.
•81. Dobb, M.G., Johnson, D.J., and Saville, B.P. (1977) J. Polym. Sci.: Polym. Symp. *58*, 237.
82. Hudson, S.D. and Lovinger, A.J. (1993) Polymer *34*, 1123.
83. Sawyer, L.C. and Jaffe, M. (1986) J. Mat. Sci. *21*, 1897.
84. Jandt, K.D., Eng, L.M., Petermann, J. and Fuchs, H. (1992), Polymer 33, 5115.
85. Lovinger, A.J. (1983) J. Polym. Sci.: Polym. Phys. *21*, 97.
86. Gross, B. and Petermann, J. (1984) J. Mat. Sci. *19*, 105.
87. Jaballah, A., Rieck, U., and Petermann, J. (1990) J. Mat. Sci. *25*, 3105.
88. Lotz, B. and Wittmann, J.C. (1984) Makromol. Chem. *185*, 2043.
89. Petermann, J., Xu, Y., Loos, J., and Yang, D. (1992) Makromol. Chem. *193*, 611.
90. Pennings, A.J., van der Mark, J.M.A.A., and Booij, H.C. (1970) Kolloid Z. u. Z. Polym. *236*, 99.
91. Chanzy, H.D., Dube, M., Marchessault, R.H., and Revol, J.-F. (1979) Biopolymers *18*, 887.
92. Gohil, R.M. and Petermann, J. (1982) Colloid and Polymer Science *260*, 312.

4   Liquid Crystalline Morphologies

4.1   Polymers Forming Liquid Crystalline Mesophases

Most homopolymers and statistical copolymers with individual flexible chains form mobile isotropic fluids over the high part of the temperature range in which the chains are stable. In an unstressed state, each chain molecule in such a fluid adopts a random arrangement of the chain units with some entanglement of adjacent coils expected. Optical or electron microscopy would be expected to show no ordered supramolecular structures. Some polymer fluids exhibit one- or two-dimensional order involving all of the chain, parts of the main chain, parts of side-chains, or parts of both the main- and side-chains over limited temperature ranges.

The ability of a polymer to form lyotropic or thermotropic liquid crystalline mesophases depends on the chemical structure. Many times the presence of flat, inflexible groupings of atoms in the main- or side-chain, with or without optically active centers, leads to liquid crystalline mesophase formation. Thermotropic mesophases are found for: 1) aromatic polyesters and polyamides with and without flexible chain components, 2) cellulose derivatives and 3) atactic polymers formed from unsaturated monomers having side chains containing rod-like groups. Examples of 1 are listed in Table 2.2 . An example of a polymer with a side-chain mesogen is poly(4'-cyanobiphenyl-4-oxy-6-hexyl acrylate) with the repeat unit [1]:

$$-CH_2-CH-$$
$$|$$
$$O=CO(CH_2)_6O-\langle\bigcirc\rangle-\langle\bigcirc\rangle-CN$$

A liquid crystalline polymer that contains a rigidly attached side-chain has the following repeat unit [2]:

$$-CH-CH_2-$$
$$CH_3O-\langle\bigcirc\rangle-\underset{O}{\overset{\ }{C}}-O-\langle\bigcirc\rangle-O-\underset{O}{\overset{\ }{C}}-\langle\bigcirc\rangle-OCH_3$$

A repeat unit containing both a side chain and a main chain mesogen is [3]:

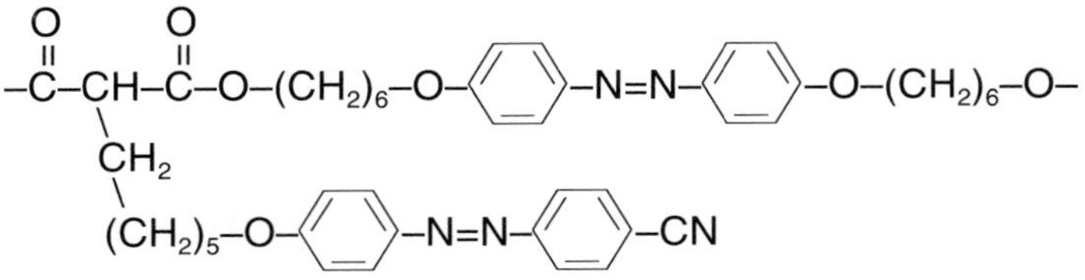

The polymers with side-chain mesogens are generally atactic and do not crystallize. However, highly isotactic or syndiotactic polymethacrylates with

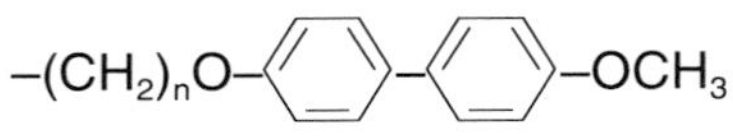

substituents have been prepared [4]. Polymers with large disk-shaped groups in the chain are not crystallizable; these include the hexakis(heptyloxy) triphenylene polymer with the repeat unit:

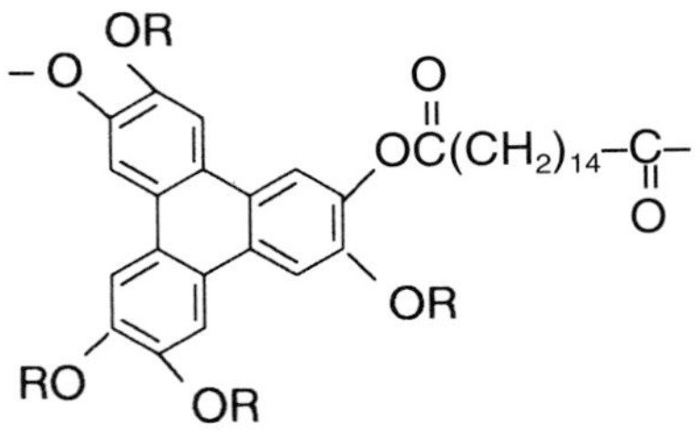

where $R = C_7H_{15}$, and polyethers with the cyclotetraveratrylene moiety:

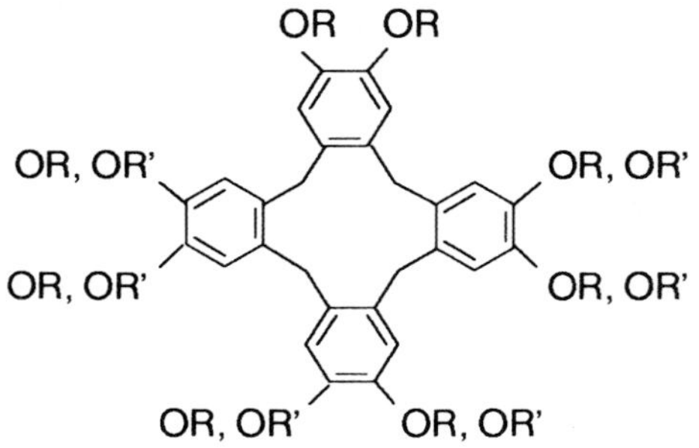

where $R = C_7H_{15}$ and $R' = C_8H_{16}$; the OR' substituents are halves of $-O(CH_2)_{16}O-$ chains linking positions on either the same or different disk-shaped groups [5, 6]. Each moiety contains two or three OR' substituents.

Some polymer chains that form liquid crystalline mesophases do so only in the presence of one or more solvents. The totally aromatic polyamides, such as poly(1,4-phenylene terephthalamide), (see Table 2.2 for structure) contain only flat, inflexible groups in the chain, as well as highly polar amide groups; therefore, these polymers do not form a liquid state in the absence of a solvent. Although they start to degrade before they melt, they do form a lyotropic liquid crystalline mesophase in an appropriate solvent. The same behavior is observed for the polybenzazoles.

The various types of liquid crystalline order and texture observed are discussed below.

4.2   Pseudohexagonal Systems

Flexible-chain polymers can exhibit mesophases in which there is side-by-side chain alignment with disorder and movement along the chain direction. One such is the pseudohexagonal or hexatic liquid crystalline state [7–9]. Adequate cooling of this mesophase can bring about crystallization and the attainment of some three dimensional order.

4.3   Rod-Like Mesogens

Polymers with chains composed of: 1) alternating rigid rod-like and flexible main chain sections ,e.g., poly(ethylene p-terphenylate) (see Table 2.2), or of 2) rigid rod-like side chain units and flexible main chain sections, e.g., poly(4'-cyanobiphenyl-4-oxyhexylacrylate), can form thermotropic liquid crystalline mesophases. Polymers having completely inflexible chains, such as polybenzothiazole, form only lyotropic mesophases. The order in a liquid crystalline mesophase of a polymer with rod-like mesogens can be one-dimensional, *nematic*, with the rigid rod-like parts of the polymer molecules possessing an average molecular orientation called the *director*. A two-dimensional arrangement, *smectic*, with the rigid sections in straight or oblique rows can also occur. A third arrangement of the rod-like sections is *cholesteric* and involves nematic layers arranged with a helical twist relative to each other. Cellulose and its derivatives are optically active and form cholesteric mesophases, as observed for 2-hydroxypropylcellulose in solution [10]. Also, a twisted smectic mesophase has been reported [11]. The cholesteric phase is only observed for optically

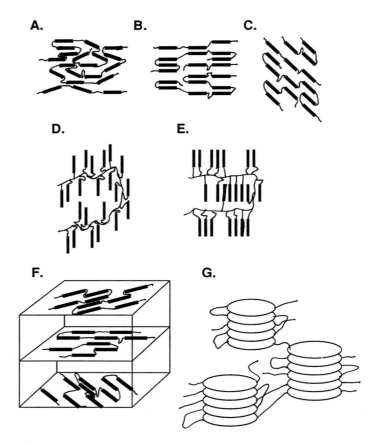

Figure 4.1   Liquid crystalline mesophases: A. main chain nematic, B. main chain smectic A, C. main chain smectic C, D. side-chain nematic, E. side-chain smectic A, F. main chain cholesteric, G. main chain discotic.

active polymers or inactive polymers mixed with optically active small molecules. Nematic, smectic and cholesteric arrangements of rod-like main-chain and side-chain polymer sections are illustrated in Fig. 4.1.

The formation of a lyotropic liquid crystalline mesophase is concentration dependent. At intermediate concentrations, the liquid crystalline mesophase will be present alone or in equilibrium with an isotropic liquid phase. At higher polymer concentrations, the liquid crystalline mesophase can be present in equilibrium with crystalline solid. The concentrations at which these phases are present depends on the polymer and the temperature.

The direct imaging of a portion of a liquid crystalline mesophase can be carried out using high resolution electron microscopy (HREM), if the phase is either stable or metastable at the measurement temperature, usually 25 °C. Such a study has been made on a smectic mesophase sample of poly(p-xylylene) [12]. Small regions of the liquid crystalline phase, stable at high temperatures, remain in a glassy liquid crystalline matrix after solution crystallization and cooling to 25 °C.

A change from one type of liquid crystalline order to a different type can occur with a change in temperature [1, 13]. The smectic mesophase is usually stable at temperatures below those at which the nematic mesophase is stable. However, exceptions to this do occur. For example, poly(4'-cyanobiphenyl-4-oxy-6 hexyl acrylate) exhibits a nematic phase between 132 °C and 124.5 °C, a smectic A mesophase between 124.5 °C and 80 °C and another nematic phase below 80 °C [1]. At temperatures above 132 °C, liquid crystalline order is not observed for this polymer and long-range disorder prevails, as is the case for any isotropic rubbery fluid. As noted above, this polymer does not crystallize; the nematic mesophase is glassy below 80 °C. The change from liquid crystal to (rubbery) isotropic liquid or from one liquid crystalline mesophase to another has the appearance of a first-order transition in volume/temperature and differential scanning calorimetry studies and involves the release of energy [11].

4.4 Disk-Shaped Mesogens

Molecules with rigid disk-shaped sections, such as the hexakis-(heptyloxy)-triphenylene polymer described above, can form *discotic* liquid crystalline mesophases in which the disk-shaped parts are stacked in columns, as illustrated in Fig. 4.1 [5, 6, 14].

4.5 Amphiphilic Polymers

Amphiphilic polymers in which alternate carbon atoms in the main chain have acidic and basic substituents have been prepared [15]. These are found to form thermotropic liquid crystalline mesophases, as a consequence of ion-dipole interaction between side-chain substituents on different chains.

4.6   Block Copolymer Mesophases

Block copolymers, containing incompatible sections, can form lyotropic liquid crystals of a different kind than polymers containing stiff units. Phase separation of AB- or ABA-type block copolymers can give spherical inclusions of one component in a continuous phase of the other, cylindrical inclusions of one component in a continuous phase of the other, or layers of the two components (see Chapter 5). (Note: in most publications the layer morphology is referred to as lamellar; in the present book this latter term will be used exclusively to identify crystallized flexible chain polymers.) When block copolymers are dispersed in certain liquid mixtures, mesophases containing a close packed array of spheres, a hexagonal array of cylinders, or a planar array of polymer layers can form [16]. This is more easily accomplished if the liquids used impart mobility to the two sequences in the block copolymer. It is expected that ABC-type block copolymers with incompatible blocks, to be discussed in Chapter 5, would show lyotropic liquid crystalline mesophases in mixed solvents. A cubic arrangement of spheres, consisting of cores of one component surrounded by a shell of a second component, in a continuous phase of the third component, a hexagonal array of cylinders, each consisting of a core of one component and a shell of a second component, in a continuous phase of the third component, or the presence of a four-layer arrangement are plausible possibilities.

4.7   Mesophase Textures

Most examinations of liquid crystalline mesophases using optical microscopy have been carried out with polarized light. For many thermotropic liquid crystalline polymeric systems, the temperature range of interest is well above room temperature, requiring the use of a hot-stage. For most types of polymer repeat units, the refractive index is different, parallel and perpendicular to the chain axis. If a number of the same type of units are aligned, the polarization of the light can be changed leading to its transmission through the analyser. A domain of rod-like mesogens oriented in the same direction can cause such a polarization change and will transmit light. Orientation of the rods perpendicular to the analyser/polarizer plane or in this plane, but parallel to either the polarizer or analyser directions, will lead to extinction of the light and the field will appear dark. Complete disorientation of the rod-like parts in the isotropic liquid also leads to extinction. Orientation of the rods in the analyser/polarizer plane, other than in the direction of the analyser or the polarizer, will lead to light being

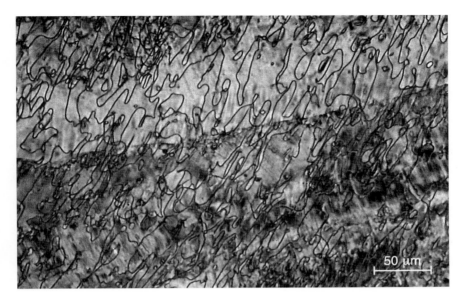

Figure 4.2   Polarizing optical micrograph of threaded liquid crystalline texture. Courtesy of C. Noel [Reprinted from: Noel, C., Friedrich, C., Laupretre, F., Billard, J., Bosio, L., and Strazielle, C. (1984) Polymer *25*, 263].

transmitted. Interaction of the light waves with the oriented mesogenic units causes a change in polarization. Since this polarization change is wavelength dependent, the light transmitted through the analyser can have one or more colors, when white light is used for the source.

Relatively thick samples of a freshly prepared liquid crystalline mesophase, viewed through crossed polaroids, appear turbid. When a nematic liquid crystalline mesophase is first formed by melting a microcrystalline film sample, many domains are present and many entangled dark lines appear in the microscopic pattern [17], as is apparent in the optical micrograph given in Fig. 4.2. Usually the number and tortuosity of these lines decreases with time due to defect coalescence and mesogen alignment. Alignment can be accelerated by shearing at low rates or by the application of an electrical or magnetic field. However, disorder will occur on cessation of shearing or removal of the field. As the temperature is increased, closed loops are formed which eventually disappear.

Thinner films of a nematic mesophase can exhibit a *Schlieren* texture; an example of this is shown in Fig. 4.3 [18]. This texture contains singular points, connected by two or four intersecting black wavy stripes or brushes, with the remaining field having one or more colors. The points are lines viewed end-on.

Figure 4.3   Polarizing optical micrograph of Schlieren liquid crystalline texture. Arrows mark singular points. Courtesy of C. Noel.

The dark brushes occur where the orientation of the director is parallel to the analyser or polarizer directions; the colored regions contain domains where the director orientation is not parallel. Upon rotating the polarizers, the brushes also rotate, either in the same direction or in the opposite direction. Preparation of nematic liquid crystalline samples by cooling very thin layers of the isotropic liquid usually leads to droplets with the Schlieren texture [19]; the droplets coalesce to form larger domains upon cooling. A Schlieren texture can contain many points randomly placed in the field.

Smectic mesophases can display a Schlieren texture with four brushes, a *focal conic* or a *fan-shaped* texture [11]. Cholesteric mesophases can show a *fingerprint* and a *ribboned* texture [10, 20, 21]. Cholesteric mesophases in reflected white light show iridescent colors [10]. For low molecular weight molecules that form cholesteric liquid crystalline mesophases with a low helical pitch, textures similar to those found for nematics appear; when the pitch is large, smectic-like textures are observed [11]. *Dendritic* and fan-shaped textures have been reported for polymers with disk-shaped mesogens [6]. The reader is referred to the *Atlas of Polymer Morphology* [22] for color micrographs showing nematic, smectic and cholesteric textures.

4.8   Mesophase Alignment

Polymers with anisotropic dielectric constants and magnetic susceptibilities are expected to be affected by the presence of electrical or magnetic fields, causing changes in the morphology. Molecules can be aligned along the applied field (positive anisotropy) or perpendicular to the field direction (negative anisotropy). Molecular alignment is greatly enhanced by the cooperativity present in liquid crystalline mesophases, crystallized systems and phase separated blends and block copolymers.

Liquid crystalline mesophases can be aligned into a single domain in an electric or magnetic field. At low magnetic field strengths the alignment of a nematic copolyester mesophase is not complete and the measured magnetic susceptibility is relatively high [11]. Increasing the field strength increases the alignment and a decrease in susceptibility occurs. Alignment of side-chain mesogens in a magnetic field have also been accomplished. At intermediate magnetic field strengths, defect clusters and inversion walls, appearing as threads which loop or terminate at a disclination, are observed for semi-flexible polyesters containing main-chain mesogens [23].

Alignment of homogenized acid-hydrolyzed cellulose microcrystals, perpendicular to the field direction, has been carried out using a 7 T magnetic field [24].

Drawing of a nematic liquid crystalline mesophase causes elongational flow which aligns the director along the streamlines [23]; on the other hand, shearing of the liquid crystalline mesophase produces *disclinations*, due to the introduction of molecular rotation.

A high degree of chain orientation was obtained in a thin film of the polymer containing main-chain/side-chain mesogens (see structure above) by annealing close to the temperature where the smectic mesophase changes to an isotropic melt [3]. However, edge dislocations were present in the mesophase lattice with one or more additional planes being added to form the defect. In some cases, cross-grated patterns, attributed to smectic plane overlap, were observed.

Investigation of a solidified discotic liquid crystalline polymer mesophase by high resolution electron microscopy shows the presence of grain boundaries [5].

The presence of folded chains in the nematic mesophase for random terpolymers containing the following repeat units:

where *n* was 4, 5, 6 or 7, was implied from electron diffraction results obtained at the lamellar crystal-liquid crystal transition [25].

4.9   Disclinations

4.9.1   Description

Disclinations are line rotational defects in the director field which give rise to the thread-like and Schlieren textures found in nematic liquid crystalline mesophases, as shown in Figs. 4.2 and 4.3 [11]. For a nematic liquid crystalline mesophase, the modes of distortion are splay, twist and bend. Large fluctuations in orientation take place in the disclination core, causing dark lines to appear [26]. Disclinations can be classified according to their topological strength and their character. Disclinations with strengths of $+\frac{1}{2}$, $-\frac{1}{2}$, $+1$ and -1 are found in nematic liquid crystalline mesophases [11]. The absolute value of the strength is given by one-half the number of brushes in a Schlieren pattern. The disclination strength is + if the brushes rotate in the same direction as the polarizers and – if they rotate in the opposite direction, indicating positive and negative points, respectively. A microscopic field will usually contain a number of disclinations with + and – strengths. Disclinations interact according to the rules of electrostatics: likes repel and opposites attract.

4.9.2   Imaging the Director Field Near a Disclination

The director field in the vicinity of a disclination in a thin sample of a nematic liquid crystalline mesophase depends on the disclination strength, and on the splay and bend elastic constants. The transmission electron microscropy (TEM) image of the director field about a disclination in a crystallizable nematic liquid crystalline polymer with a disclination strength of $+\frac{1}{2}$ is given in Fig. 4.4 [26]. To obtain this image, the mobile nematic mesophase was first quenched to form a nematic glass; then, the temperature was raised to about 20 °C below the crystalline melting point and the glass annealed, leading to the formation of many lamellar crystals with sizes smaller than the disclination size [27]. The lamellas are viewed edge-on and the director is normal to them. The many striations observed are composed of lamellas. The crystals are oriented with respect to one another in accord with the orientation in the supercooled nematic mesophase. For the particular sample imaged in Fig. 4.4, the director field

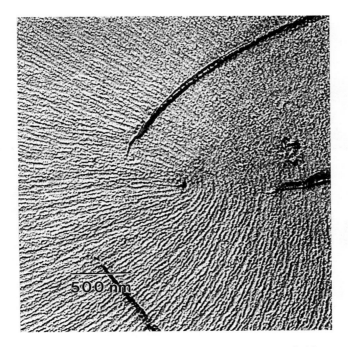

Figure 4.4   Transmission electron micrograph of a director field near a s = $+^1/_2$ disclination in a nematic liquid crystalline mesophase. Courtesy of S.D. Hudson [Hudson, S.D. and Thomas, E.L. (1991) Chemtracts–Macromolecular Chemistry *2*, 73. Copyright © 1991, Data Trace Publishers, Inc. Published in Chemtracts–Macromolecular Chemistry *2* (2) as part of an article titled "Disclinations and Inversio Walls in Nematic LPC's," pp. 73–93, and reproduced here with permission].

generally curves around the disclination at the center of the photograph. The heavy dark lines are cracks occurring parallel to the director field direction.

A TEM image of the director field in the vicinity of disclinations in a smectic mesophase found for an aromatic polyester is shown in Fig. 4.5 [28]. Focal conic disclinations are revealed in this photograph. These parabolic disclinations occur in pairs which are mutually orthogonal and with the vertex (turning point) of one intersecting the focus of the other. Smectic-A mesophases have been imaged for poly(ester imide)s [29].

4.9.3   Disclination Clusters

Disclinations with + and − strengths can occur in dipolar or in quadrapolar (Lehman) clusters. Nematic polymer mesophases aligned in a magnetic field of

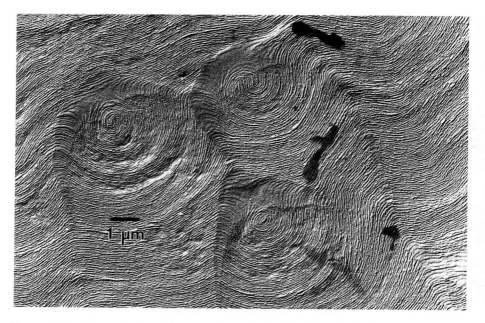

Figure 4.5   Transmission electron micrograph of a director field near focal conic disclinations in a smectic liquid crystalline mesophase. Courtesy of S.D. Hudson [Reprinted with permission from: Hudson, S.D., Lovinger, A.J., Larson, R.G., Davis, D.D., Garay, R.O. and Fujishiro, K. (1993) *Macromols.* *26*, 5643. Copyright © 1993, American Chemical Society].

intermediate field strength display both dipolar and quadrapolar clusters, and extensional flow of oriented nematic liquid crystals brings about the appearance of many Lehman clusters [23].

4.9.4   Inversion Walls

Inversion walls, where the director inverts by 180° from one side to the other, have been observed in nematic liquid crystalline mesophases under the same conditions which produce Lehman clusters. Extensional flow by surface tension spreading of a thermotropic nematic liquid crystalline mesophase on hot H_3PO_4 brings about the appearance of quadrapole (Lehman) clusters and inversion walls. Inversion walls can be imaged using the lamellar decoration technique and TEM, as discussed above for the imaging of the director field near a disclination.

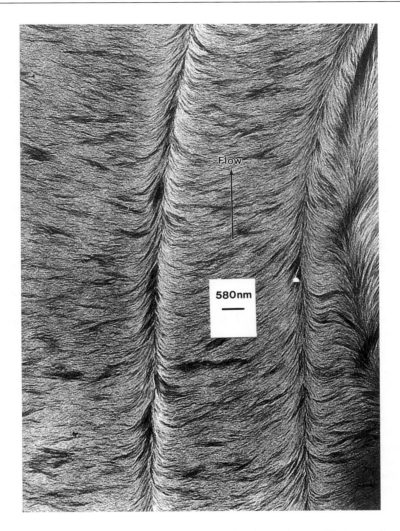

Figure 4.6   Transmission electron micrograph of an inversion wall in extensional flow of a nematic liquid crystalline mesophase. Courtesy of S.D. Hudson [Reprinted from: Hudson, S.D. and Thomas, E.L. (1991) Phys. Rev. A *44*, 8128. Copyright © 1991, The American Physical Society].

A TEM photograph of an inversion wall for a nematic liquid crystalline polymer mesophase formed by pouring the oriented mesophase onto a cold surface is shown in Fig. 4.6 [23].

References

1. Le Barny,P., Dubois, J.C., Friedrich, C. and Noel, C. (1986) Polym. Bull. *15*, 341
2. Xu, G., Wu, W., Shen, D., Hou, J., Zhang, S., Xu, M. and Zhou, Q. (1993) Polymer *34*, 1818.
3. Voigt-Martin, I.G. and Durst, H. (1989) Macromols. *22*, 168.
4. Nakano, T., Hasegawa, T. and Okamoto, Y. (1993) Macromols. *26*, 5494.
5. Voigt-Martin, I.G., Garbella, R.W. and Schumacher, M. (1992) Macromols. *25*, 961.
6. Percec, V., Cho, C.G., Pugh, C. and Thomazos, D. (1992) Macromols. *25*, 1164.
7. Kojima, M. and Magill, J.H. (1986) J. Mat. Sci. *21*, 2651.
8. Tadokoro, H. (1979) *Structure of Crystalline Polymers*. New York: Wiley.
9. Wunderlich, B. (1973) *Macromolecular Physics, Vol. 1*. New York: Academic Press.
10. Werbowyj, R.S. and Gray, D.G. (1976), Mol. Cryst. Liq. Cryst. Lett. 34, 97.
11. Noel,C. (1985) in *Polymer Liquid Crystals*, A. Blumstein, ed. New York: Plenum Publishing Corp., p. 21.
12. Zhang, W. and Thomas, E.L.(1992) J. Polym. Sci.: Polym. Phys. *30*, 1285.
13. Meurisse, P., Noel, C., Monnerie, L. and Fayolle, B. (1981) Brit. Polym. J. *13*, 55.
14. Kreuder, W., Ringsdorf, H. and Tscherner, P. (1985) Makromol. Chem. Rapid Commun. *6*, 367.
15. Paleos, C.M., Tsiourvas, D., Anaslassopoulou, J. and Theophanides, T. (1992) Polymer *33*, 4047.
16. Wittmann, J.C., Lotz, B., Candau, F. and Kovacs, A.J. (1982) J. Polym. Sci.: Polym. Phys. *20*, 1341.
17. Noel, C., Friedrich, C., Laupretre, Billard, J., Bosio, L. and Strazielle, C. (1984) Polymer *25*, 263.
18. Noel, C. (1988) in *Atlas of Polymer Morphology*. Munich: Hanser Publishers, pp. 219, 229.
19. Galli, G., Laus, M., Angeloni, A.S., Ferruti, P. and Chiellini, E. (1985) Eur. Polym. J. *21*, 727.
20. Werbowyj, R.S. and Gray, D.G. (1984), Macromols. 17, 1512.
21. Galli, G., Chiellini, E. and Angeloni, A.S. (1988) in *Atlas of Polymer Morphology*. Munich: Hanser Publishers, pp.219, 231, 233.
22. Woodward, A.E. (1988) *Atlas of Polymer Morphology*. Munich: Hanser Publishers.
23. Hudson, S.D. and Thomas, E.L. (1991) Phys. Rev. A *44*, 8128.
24. Sugiyama, J., Chanzy, H. and Maret, G. (1992) Macromols. *25*, 4232.
25. Kent, S.L. and Geil, P.H. (1992) J. Polym. Sci.: Polym. Phys. Ed. *30*, 1489.
26. Hudson, S.D. and Thomas, E.L. (1991) Chemtracts-Macromol. Chem. *2*, 73.
27. Wood, B.A. and Thomas, E.L. (1986) Nature (London) *324*, 655.
28. Hudson, S.D., Lovinger, A.J., Larson, R.G., Davis, D.D., Garay, R.O. and Fujishiro, K. (1993) Macromols. *26*, 5643.
29. Parley, R., Shen, D., Gabori, P.A., Harris, F.W., Cheng, S.Z.D., Adduci, J., Facinelli, J.V. and Lenz, R.W. (1993) Macromols. *26*, 3687.

5 Morphologies of Blends, Block Copolymers, Composites and Laminates

5.1   Polymer Compatibility

The compatibility of two polymers in mobile liquid states depends mainly on the forces acting between the various groups in the chains of the same material as well as between the groups in the chains of the two different materials. In nonpolar or weakly polar polymers, the physical forces acting are principally dispersion forces. Generally speaking, the attractive forces between different kinds of nonpolar chains are not of sufficient strength to bring about compatibility. Therefore, when two nonpolar polymers in mobile isotropic liquid states are mixed together in a blend, phase separation usually occurs. If sections of these two chains are covalently linked, forming a block copolymer, microphase separation will take place. In a triblock copolymer the compatibility will vary between the three pairs, affecting the morphology. The compatibility of two polymers will be concentration, temperature and molecular weight dependent. It is possible, using quenching techniques, to obtain a non-equilibrium degree of order in a two-component system and thereby achieve compatibility.

Polar polymers reported to be miscible include: 1) statistical copolymers of styrene and acrylonitrile blended with atactic poly(methyl methacrylate), PMMA, 2) poly(vinylidene chloride) blended with PMMA and, 3) atactic polystyrene blended with poly (p-phenylene oxide) [1]. Polar groups in these polymers are:

$\overrightarrow{C\equiv N}$ in polyacrylonitrile,

$\overset{O\nearrow}{\underset{C-O}{\diagup}}$ in PMMA ,

$\overset{Cl}{\underset{C-Cl}{|}}\nearrow$ in poly(vinylidene chloride), and

$\overset{O}{\underset{C \uparrow C}{\diagdown}}$ in poly(p-phenylene oxide).

The (+) and (–) parts of the dipole are marked by an arrow pointing from (+) to (–). The (+) part of one dipole would attract the (–) part of another dipole on the same or different chain.

Some examples of polymer pairs that phase separate are: polystyrene with one of the following: polybutadiene, polyisoprene, poly(dimethyl siloxane), an ethylene/butene statistical copolymer, poly(4-vinyl pyridine) or poly(2-vinyl pyridine) and the pair polybutadiene/poly(dimethyl siloxane). (Repeat unit structures are given below or in Table 2.1.) Polystyrene is weakly incompatible with poly(methyl methacrylate).

$$-CH_2-CH- \qquad -CH_2-CH- \qquad \begin{array}{c} CH_3 \\ | \\ O \\ \diagdown \\ -O-Si-O- \\ \diagup \\ O \\ | \\ CH_3 \end{array}$$

4-vinyl pyridine   2-vinyl pyridine   dimethylsiloxane

If the polymers in the blend or the chain sections in the block copolymer crystallize upon cooling from the mobile liquid, then additional phases are formed. A number of AB, ABA and ABC type block copolymers are prepared from unsaturated monomers. The crystallizability of chain segments in these copolymers will be the same as the parent atactic homopolymer; polybutadiene or polyisoprene units present in these block copolymers are a mixture of isomers and, therefore, do not crystallize. Repeat units that occur are:

$$\overset{R}{\underset{-CH_2-C=CH-CH_2-}{|}} \qquad \overset{CH=CH_2}{\underset{-CH_2-CR-}{|}}$$

cis or trans 1,4-   1,2-

polydiene repeat units

where R is CH_3 in polyisoprene and H in polybutadiene. An ethylene/butene copolymer can be formed by hydrogenation of isomeric polybutadiene.

Some AB block copolymers have been prepared by sequential polymerization of a ring monomer and an unsaturated monomer, or of two ring monomers. An example of the former is the diblock copolymer containing poly(ethylene oxide) and atactic polystyrene, and an example of the latter is the diblock copolymer containing poly(ε-caprolactam) and poly(dimethyl siloxane). The poly(ethylene oxide) and the poly(ε-caprolactam) portions of these block copolymers are crystallizable.

The types of systems giving distinct phase separated morphologies to be discussed herein include: 1) blends of two noncrystallizing polymers, 2) noncrystallizing di- and tri-block copolymers, 3) binary and ternary noncrystallizing blends containing a block copolymer, 4) binary blends containing one and two crystallizable polymers, 5) crystallizable diblock copolymers, and 6) segmented block copolymers containing crystallizable segments. Composites and laminates are additional multiphase systems with distinct form and structure.

5.2   Binary Blends of Noncrystallizing Homopolymers

Mixing of two incompatible amorphous homopolymers sufficiently, prior to obtaining a rigid or semi-rigid mass, can lead to microphase separation, in which particles of the constituent present in the smaller quantity are dispersed in the other, more prevalent, constituent. If the two constituents are partially miscible at the mixing temperature, then upon cooling the mixture, subinclusions of the dominant constituent within the dispersed particles of the other component can also appear. Cooling of the mixture to form a glass of one or both components gives a metastable system; at equilibrium, two separated phases would be present, one rich in one component and the other rich in the additional component. The size of the dispersed particles depends on the polymers chosen, the temperature and the mixing speed, as well as other processing variables. The dispersed particles in an unstressed blend are expected to be approximately a spherical collection of randomly coiled chains. A spherical inclusion would minimize the surface area present and the interfacial energy. The application of mechanical stress during mixing could lead to elongated dispersed particles.

Various electron microscopic investigations, scanning electron microscopy (SEM) and transmission electron microscopy (TEM), have been carried out at 25 °C on amorphous polymer blends containing two polymers both with T_g's well above 25 °C [2–4]. Blends containing a rubbery polymer at 25 °C dispersed in a glassy one have also received attention [5, 6]. Fracture surfaces and microtomed sections of thick samples as well as the surfaces of thin films have been studied. Reactions with reagents containing heavy metal atoms have been used to preferentially stain one of the components. SEM and TEM studies of fracture surfaces of amorphous blends usually, but not always, show spherical or semi-spherical inclusions. The subinclusions of the continuous component observed within the dispersed component particles are found to be spherical.

The application of an electrical field (strengths of 1000–3000 V/cm) to a poly(methyl methacrylate)/polystyrene/toluene mixture causes the formation of ellipsoidal or columnar structures [7]; removal of the field in the presence of a considerable amount of toluene gives spherical inclusions.

5.3   Noncrystallizing AB- and ABA-Type Block Copolymers

Phase separation in noncrystallizing block copolymers display features not found for blends. This is caused by restrictions on the phase separated regions in block copolymers as a consequence of the covalent attachment of a sequence of one type of repeat unit to a sequence of another type. This attachment causes the two different chain portions to stay in close proximity to each other, greatly limiting the size of the separated phases and also affecting the morphology of the phase separated system. The phase separated morphologies observed for two component block copolymer systems (AB, ABA and star block) include: 1) layers of A separated by layers of B [8], 2) spheres of A embedded in a continuous B phase [9], 3) cylinders of A in a continuous B phase [10], and 4) the bicontinuous double diamond type formed by star blocks [11]. (Note: The layer morphology is referred to as lamellar in many publications and books. In the present book, lamellar will be a term reserved for crystallizing systems.) The arrangements found for two component block copolymers are shown schematically in Fig. 5.1. The precise morphology adopted by a phase separating AB-type block copolymer depends on the composition, changing from dispersed spheres to dispersed cylinders to the ordered bicontinuous double diamond arrangement to layers with increasing volume fraction of the dispersed component [12–14]. The miscibility of the two blocks in noncrystallizing AB block copolymers of polystyrene and polyisoprene in toluene was shown to depend on the molecular weight [15]. Morphological changes were found for noncrystallizing diblock copolymers of poly(dimethyl siloxane) (PDMS) and polybutadiene with a change in the total molecular weight [13]; an increase in the overall molecular weight caused a change from rod-like to spherical inclusions. A change in

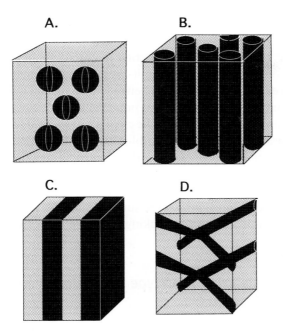

Figure 5.1   Microphase separation in AB-type block copolymers: A. spherical inclusions, B. cylindrical inclusions, C. layers, D. ordered bicontinuous double diamond inclusions.

morphology from layers to dispersed spheres was found for an ABA block copolymer system when the A component was interchanged with the B component [16]. Heating amorphous block copolymers cast from solution to a temperature above T_g for all of the components present can lead to morphological changes. When a methyl ethyl ketone-cast film of a triblock copolymer of polystyrene/polybutadiene/polystyrene, containing 56 wt % polystyrene, was annealed at 150 °C, the morphology changed from cylinders of polybutadiene in a continuous matrix of polystyrene to alternating layers of the two polymers [17].

Microdomain formation in amorphous block copolymers has been observed in solution, in films cast from solution, in rubbery liquids and in glassy liquids. The layer thickness and the diameter of the spherical or cylindrical domains formed depends on the block length and is usually in the 10–20 nm range.

TEM is used on thin sections of samples to observe separated particles of the size found in block copolymers. In thin sections, spherical inclusions will appear as circles of different sizes depending on the positions of the spheres; cylindrical inclusions will appear as circular, ellipsoidal or as layer-like depending on the direction of the slice. In order to distinguish spherical from cylindrical inclusions and cylindrical inclusions from layers in a particular preparation, it is necessary to view sections cut in different directions through the sample.

Many of the block copolymers studied to date contain isomeric polybutadiene units or isomeric polyisoprene units as one of the blocks; regions containing these can be easily distinguished when reacted with OsO_4. RuO_4 is employed as a staining agent for polystyrene and polyethylene blocks.

AFM has been used to investigate the changes of surface roughness with time for a phase-separated amorphous AB diblock copolymer [18]. AB block copolymers with a layer morphology show surface corrugations that have (+) and (–) type disclinations [19]. Regions of intersecting layers of the two components are seen, resulting in circular elevated patches of one of the components. A two-dimensional array of depressions at the surface is apparent for samples containing spherical domains, due to the packing of the spheres close to the surface.

5.4   Noncrystallizing ABC-Type Block Copolymers

For three component block copolymer systems (ABC-type with B the middle block) the phase separated morphologies first reported were: spheres of A surrounded by shells of B in a continuous C phase (Fig. 5.2A), cylinders of A surrounded by cylindrical shells of B in a continuous C phase (Fig. 5.2C) and layers of A, B and C in the order A - B- C - B (Fig. 5.2F) [20, 21].

Seven additional morphologies have been recently found for ABC block copolymers. One of these contains spheres of A, surrounded by hexagonal-shaped shells of B in a continuous C phase [22].

Two morphologies have been reported for a triblock copolymer containing one component that is strongly incompatible with the other two [23]; in one of these arrangements, the strongly incompatible B component forms cylinders at the interface between layers of the A and C components (Fig. 5.2G) and in the other isolated rings of the strongly incompatible B component exist on cylinders of the A component in a matrix of the C component (Fig. 5.2D). TEM micrographs showing these latter two morphologies are given in Fig. 5.3 and 5.4. In these two cases the A blocks are polystyrene, the B blocks are an ethylene/butene copolymer and the C blocks are poly(methyl methacrylate). Reaction with RuO_4 causes the polystyrene and the ethylene/butene copolymer regions to appear dark and the poly(methyl methacrylate) portions white. (In Fig. 5.4 a cross section of one of many rings of the B block component around cylinders of the A block component is marked by an arrow.) Changing the B component leads to

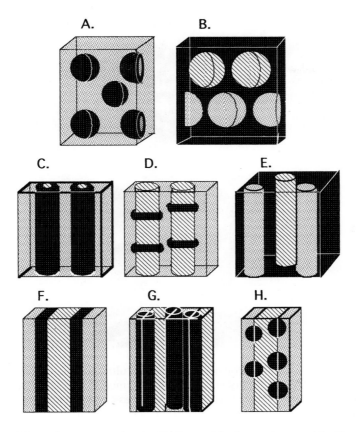

Figure 5.2   Microphase separation in ABC-type block copolymers (the B portion is in black): A. spherical inclusions with spherical shells, B. separate spherical inclusions of A and C in B, C. cylindrical inclusions with cylindrical shells, D. cylindrical inclusions with rings, E. separate cylinders of A and C in B, F. four layers, G. cylindrical inclusions at layer interfaces, H. spheres at layer interfaces.

a third arrangement with balls of that component at the interface of layers of the A and C components (Fig. 5.2H) [24]; a TEM micrograph demonstrating this morphology when the B block is polybutadiene is given in Fig. 5.5. Reaction with OsO$_4$ gives dark polybutadiene regions and white poly(methyl methacrylate) domains; due to partial miscibility of the low molecular weight polybutadiene chain sections with the polystyrene portions, this component appears gray.

Another morphology reported for an ABC triblock copolymer is the ordered tricontinuous double diamond type, containing double diamond frameworks of the A and C regions that are mutually interpenetrating in a continuous matrix of

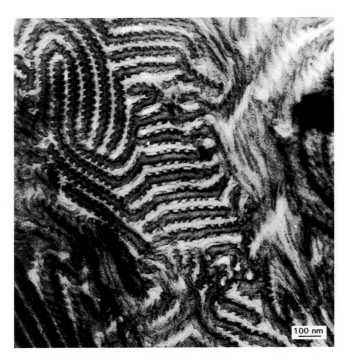

Figure 5.3   Transmission electron micrograph of an ultrathin microtomed section of a 48 wt % polystyrene/17 wt % ethylene/butene copolymer/35 wt % poly(methyl methacrylate) block copolymer reacted with RuO_4. Courtesy of R. Stadler [Reprinted with permission from: Auschra, C. and Stadler, R. (1993) Macromols. *26*, 2171. Copyright © 1993, American Chemical Society].

B [25]; this morphology was obtained for a solution cast ABC triblock copolymer of polyisoprene (A block), polystyrene (B block) and poly(2-vinyl pyridine) (C block) with equal sized A and C sections and a B section having a volume fraction of 0.48 to 0.66. A TEM micrograph showing this morphology is given in Fig. 5.6. Due to OsO_4 staining and thickness variations, one strut of the polyisoprene framework appears black; the other polyisoprene struts and one of the poly(2-vinyl pyridine) (P2VP) struts appear gray; while the remaining P2VP struts and the polystyrene portions are white. This arrangement is a stable one which does not change upon annealing. Changing the volume fraction of the B component while keeping the volume fractions for the A and C components equal leads to the formation of three other morphologies; these are the layer type, interpenetrating lattices of cylinders of the A and C chain blocks in a continuous B matrix (Fig. 5.2E), and interpenetrating lattices of spheres of A and C in a continuous matrix of B (Fig. 5.2B) [14].

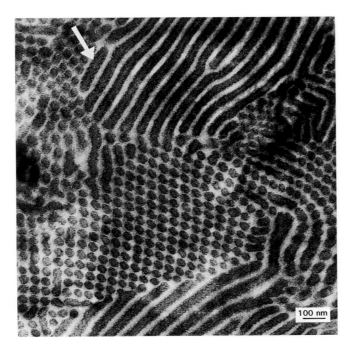

Figure 5.4   Transmission electron micrograph of an ultrathin microtomed sectin of a 45 wt % polystyrene/6 wt % ethylene/butene copolymer/49 wt % poly(methyl methacrylate) block co-polymer reacted with RuO_4. Courtesy of R. Stadler [Reprinted with permission from: Auschra, C. and Stadler, R. (1993) Macromols. *26*, 2171. Copyright © 1993, American Chemical Society].

The morphologies of ABC triblock copolymers depend on the amount of the center (B) component [14, 23]. For the triblock copolymer containing polystyrene (A component), a statistical copolymer of ethylene and butene in the center (B component) and poly(methyl methacrylate) (PMMA) (C component) at the highest statistical copolymer content used (38 wt %), the four-layer morphology appeared. This changed to cylinders of the statistical copolymer at the interfaces of layers of the other two components with a decrease in copolymer content (17%), and, with a further decrease (6%), cylinders of copolymer with isolated rings of polystyrene in a PMMA matrix were observed [23]. For ABC block copolymers containing polyisoprene (A), polystyrene (B), and poly(2-vinyl pyridine) (C), the morphologies appearing are: 1) spheres of the A and C components in B (>80 vol % B), 2) cylinders of A and C in B (68–76 vol% B), 3) ordered tricontinuous double diamond arrangement of A and C in B (50–65 vol% B), and 4) the four-layer arrangement (30–40 vol% B) [14].

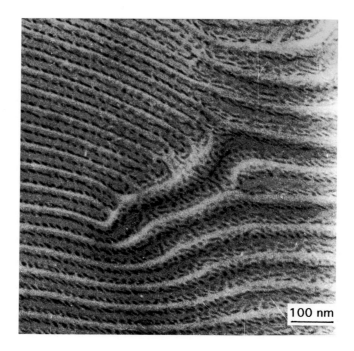

Figure 5.5   Transmission electron micrograph of an ultrathin microtomed sectin of a 45 wt % polystyrene/6 wt % polybutadiene/49 wt % poly(methyl methacrylate) block copolymer reacted with OsO$_4$. Courtesy of R. Stadler [Reprinted with permission from: Bechmann, J., Auschra, C. and Stadler, R. (1993) Makromol. Chem. Rapid Commun. *15*, 67. Published by Huthig & Wepf Verlag, Basel, Switzerland].

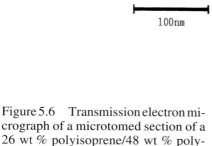

100nm

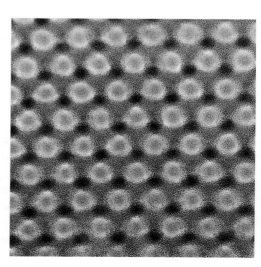

Figure 5.6   Transmission electron micrograph of a microtomed section of a 26 wt % polyisoprene/48 wt % polystyrene/26 wt % poly(2-vinyl pyridine) block copolymer reacted with OsO$_4$. Courtesy of Y. Matsushita [Reprinted with permission from: Mogi, Y., Mori, K.,Matsushita, Y. and Noda, I. (1992) Macromols.*25*, 5412. Copyright © 1992, American Chemical Society.

5.5 Binary Blends Containing a Noncrystallizing Block Copolymer

The morphology of blends containing particles of an amorphous homopolymer dispersed in a continuous matrix of a second amorphous homopolymer was discussed in section 5.2. When a block copolymer is used in place of one of the homopolymers, the morphology is expected to differ from that found for homopolymer blends. If the block copolymer constituents are incompatible and the block copolymer is not compatible with the homopolymer, then two types of morphology are expected, depending on composition. One of these would consist of rounded particles of the microphase separated block copolymer in a continuous homopolymer matrix [26], and the other would involve rounded homopolymer particles in a microphase separated block copolymer matrix. Subinclusions might possibly be observed.

Additional morphologies, as well as these, have been reported. One consists of small interspersed regions of each constituent [26]; a second type, found for a blend containing an AB block copolymer, is the double diamond structure, a morphology associated with star block copolymers alone [27, 28]. A third morphology exhibited by block copolymer/homopolymer blends consists of lamellar channels connected by periodic orthogonal columns, which is a mesh or lamellar cantenoid structure and is found for AB and star block types [27, 29]. In AB block copolymer/homopolymer blends having the double diamond and the lamellar cantenoid structures, some phase mixing was observed [27]. TEM micrographs of thin sections of blends containing a polystyrene/polyisoprene AB block copolymer and polystyrene, reacted with OsO_4 are shown in Fig. 5.7 [27]. The two blocks in the copolymer have the same molecular weight, 15,000. The morphology observed depends on the molecular weight of the atactic polystyrene component, with the lamellar cantenoid type occurring at M = 6500 (Fig. 5.7A and 5.7B) and the ordered bicontinuous double diamond type present when M = 3000 (Fig. 5.7C).

5.6 Ternary Blends Containing a Noncrystallizing Block Copolymer

Ternary blends of a noncrystallizing diblock copolymer with the two related homopolymers have been prepared and observed using TEM; rounded particles containing portions made up of concentric shells of the block copolymer and portions of one of the constituent homopolymers were found [26].

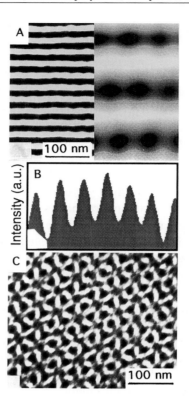

Figure 5.7   Transmission electron micrographs of blends of an equimolar polyisoprene/polystyrene block copolymer (M = 30,000) with atactic polystyrene; thin sections reacted with OsO$_4$ were viewed: A. M(PS) = 6500; the enlargement was 2D Fourier filtered, B. trace through a polyisoprene-rich lamella, C. M(PS) = 3000. Courtesy of R.J. Spontak [Reprinted from: Spontak, R.J., Smith, S.S. and Asraf, A. (1993) Polymer Comm. *34*, 2233].

5.7   Blends Containing One Crystallizable Polymer

If one of the constituents in a binary blend of two flexible homopolymers is crystallizable, spherulites or hedrites can be prepared either from the melt or from a cast film [30–33].

If the two polymers are miscible, extinction rings, as described in Chapter 3, can appear on the spherulites where none exist for the pure polymer [31]. An example of this is shown in Fig. 5.8 which contains polarizing optical micrographs of poly(ε-caprolactam) alone and blended with 10% of a styrene/acrylo-

Figure 5.8   Polarizing optical micrograph of poly(ε-caprolactone) (PCL) spherulites crystallized from a blend with a styrene/acrylinitrile copolymer: A. 100% PCL, B. 90% PCL. Courtesy of W. Li [Reprinted from: Li, W., Yan, R. and Jiang, B. (1992) Polymer *33*, 889].

nitrile copolymer. As the copolymer content is increased, the spherulites lose their symmetry and the banding becomes more irregular. It is believed that the rigidity of the copolymer chains, which are residing only in the amorphous interlayers of the spherulites, causes lamellar twisting, observable as extinction rings.

The effects of blending depend on the composition [32]; an 80/20 wt % blend of poly(ethylene oxide) and poly(methyl methacrylate) show only spherulites of the former under the optical microscope while a 50/50 blend exhibits both spherulitic and noncrystalline regions coexisting [32]. Blends of two polymers with the same repeat unit but different chain configurations, such as atactic and isotactic polystyrene, can exhibit a high degree of incompatibility with isolated hedrites of the isotactic polymer being observed [33].

The morphology can change with the crystallization temperature, as found for blends of crystallizable poly(ether ether ketone) and noncrystallizable polyetherimide [30]. The noncrystallizable component appears between spherulites at high T_c and in pockets between lamellar bundles within spherulites at lower T_c; at still lower T_c, the crystallizing component forms short faceted lamellas in dendrites.

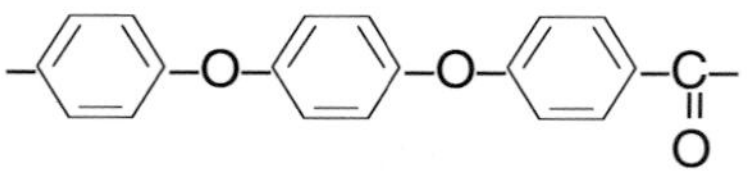

poly(ether ether ketone)

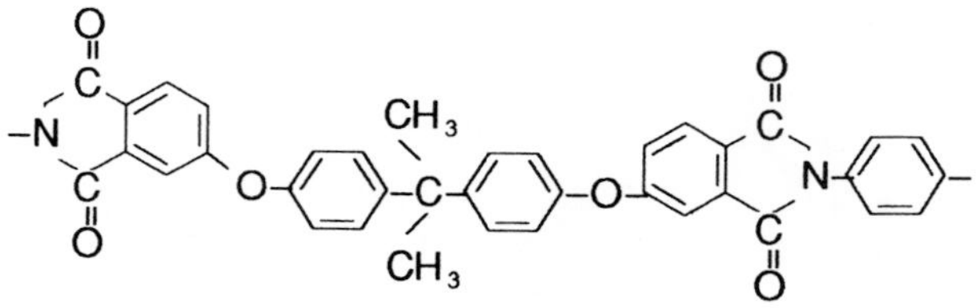

polyetherimide

5.8   Blends Containing Two Crystallizable Polymers

Crystallization from the melt of two crystallizable homopolymers, such as polyethylene and polypropylene, which do not co-crystallize, can lead to spherulites of one component mixed with spherulites of the other [34]. In blends of these two polymers the spherulite size for polypropylene is greatly reduced by the presence of small amounts of polyethylene.

The amount of crystallization of one of the components in a binary blend can be affected by chemical changes in the other component [35]; for example, the crystallization of poly(ethylene oxide) is abruptly decreased in the presence of more than 70% hydroxypropylcellulose but is not affected by this amount of carboxymethylcellulose.

Elongated growths on spherulites and banded spherulites with a change in the banding with growth have been observed in blends of polyhydroxbutyrate with a copolymer of hydroxybutyrate:

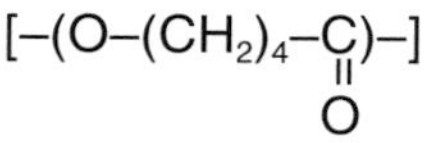

$$[-(O-(CH_2)_4-C)-]$$
$$\underset{O}{\|}$$

and hydroxyvalerate [36]:

$$[-(O-(CH_2)_5-\underset{\underset{O}{\|}}{C})-]$$

Blends of two crystallizable polymers, such as linear and (short-chain) branched polyethylene, can show a progressive change from curved to straight lamellas accompanied by a decreasing average amorphous thickness as the composition is changed [37].

5.9   Crystallizable AB and ABA Block Copolymers

Amorphous AB and ABA type block copolymers with incompatible components microphase separate, giving spherical, cylindrical or layer morphologies, as discussed in section 5.3. Flexible chain block copolymers containing a crystallizable component would not have these morphologies. For quiescently crystallized flexible chain AB or ABA block copolymers, single lamellas or multilamellar structures would appear depending on the conditions. Prenucleated crystallization from dilute solution is expected to give single lamellas [38]. In these lamellas most of the crystal stems would be connected by folds containing units of the same component; for diblock copolymers one of the chain ends would mainly contain repeat units of the noncrystallizable component. If the two components are incompatible, the end section will separate from the folds. If the noncrystallizable component is large, separate microphases of it will form near the lamellar surface. Hedrites are expected if crystallization occurs from moderately concentrated solution without seeding. Parts of the noncrystallizing chain ends, particularly for those on chains in the middle of the lamellar ribbon, will be trapped between the lamellas in the hedrite.

ABA block copolymers with crystallized poly(ε-caprolactam) A blocks and a noncrystallizable (at 25 °C) polydimethysiloxane B block have been obtained from the melt [39]. The morphology observed depends on the molecular weight of the A blocks. Spherulites appear, if the A block molecular weight is high enough to permit chain folding ; at a lower A block molecular weight, axialites were seen. Spherulites of another ABA block copolymer with crystallizable A blocks (poly(ethylene oxide) and a noncrystallizable B block (atactic polystyrene) have been obtained by casting from solution [40]. The morphology and crystallinity obtained is dependent on the casting solvent. When the solvent is a good one for both components, rather than preferring the crystallizing component, the crystallinity is lower and the spherulites are small and ill-defined.

Blends containing an AB block copolymer having a crystallizable component have received little attention to date. However, solvent casting of a blend of polystyrene with a poly(ethylene oxide)/polystyrene block copolymer in the presence of an electric field leads to pearl-chain spheres or long columnar structures [41]; at higher volume fractions of poly(ethylene oxide), acting as the dispersed component, the columnar inclusions were observed.

ABA block copolymers have been prepared with poly(γ-benzyl-L-glutamate):

$$[-NH-\underset{R}{CH}-\underset{O}{C}-]$$

where R = CH$_2$–CH$_2$–C(=O)–O–CH$_2$–⟨◯⟩ as the A part and atactic polystyrene as the B part [42].

The polypeptide forms a rigid rod in the bulk and in certain liquids and the length of this portion determines the morphology.

5.10   Segmented Block Copolymers

In segmented block copolymers, the sections of one repeat unit alternate with sections of another many times per chain. Segmented block copolymers can be prepared by surface reaction of polydiene lamellas in suspension [43] or by functional group reactions in which oligomers are involved [44]. Segmented block copolymers of trans-1,4-polyisoprene (TPI) units, epoxidized TPI units, and TPI units and hydrochlorinated TPI crystallize from solution [45, 46].

epoxidized TPI   hydrochlorinated TPI

The hedrites generally observed contain TPI crystal stems, amorphous TPI and modified TPI chain sections. The crystalline fraction observed for solution-crystallized TPI block copolymers is considerably lower than that found for solution-crystallized TPI.

Segmented block copolymers, prepared via functional group reactions, usually contain rigid (hard) sections, such as polyurethane, polyamide, aromatic polyester or polydiacetylene/polyurethane and long flexible (soft) sections, such as an aliphatic polyester, polyether or polybutadiene [44, 47–49]. One example is a segmented block copolymer with the following diacetylene/urethane hard segments:

$$-[OCH_2C{=}C{-}C{=}C{-}CH_2O{-}\overset{\overset{\displaystyle O}{\|}}{C}{-}NH{-}\langle\bigcirc\rangle{-}CH_2{-}\langle\bigcirc\rangle{-}NH{-}\overset{\overset{\displaystyle O}{\|}}{C}{-}]_p-$$

and poly(ether-urethane) soft segments [47].

Condensation-type segmented block copolymers have been crystallized and studied by TEM with lamellar structures being apparent [49–52].

5.11   Composites and Laminates

Composites, heterogeneous mixtures of two or more components can be prepared with a linear-chain polymer or a rigid network as the continuous phase and short fibers or particles of another material making up the dispersed phase [53]. Both crystallizable and noncrystallizable linear-chain polymers are used to form composites. Glass fibers, glass beads, carbon fibers, carbon black, kaolin and mica are some of the particulate materials used [54–56]. More than one type of filler can be used in the same composition.

The particle size of the dispersed material is usually many times larger than that found in blends containing a spherical or near spherical dispersed phase. The shape of the dispersed particles will depend on the material used and can be spherical, rod-like or flat. The degree of orientation of fibrous components will depend on the composition and the sample preparation method; in thick samples, fiber orientation varies with sample depth. Various fiber reinforcement patterns can be used, including: uniaxially oriented, biaxially oriented and random. If a composite contains particulates, the degree of dispersion will depend on the materials and the mixing technique.

The components of a composite are incompatible and form separate phases. The dispersed component in a composite will have the same form before and after formation of the composite, although the particle size or fiber length, depending on the material, can be reduced during processing if vigorous mixing or grinding is used. Particularly during injection molding, the fiber length can be greatly reduced. The morphology of the continuous polymer phase in a composite containing particulates can differ from that for the polymer processed alone. Particulates can act as nucleating agents for crystallization and lead to a loss of spherulite texture. As noted in Chapter 3, transcrystallization occurs with certain fibers in a supercooled crystallizable polymer melt.

Laminates usually contain alternating layers of two or more materials. Laminates can be formed using pieces of a woven fabric as one component and a liquid prepolymer that polymerizes to give a rigid network polymer as the other component. Laminates can be prepared containing alternating layers of two linear chain materials or of layers of a continuous fiber/rigid network composite with a change in the fiber orientation from layer to layer [57].

To investigate composites, SEM study of fracture surfaces is frequently used. Fiber orientation, fiber size and adherence of portions of the continuous phase to the fiber or particulate surfaces (evidence of surface wetting) are of interest [56].

References

1. Sperling, L.H. (1992), *Introduction to Physical Polymer Science*, 2nd Ed. New York: Wiley-Interscience.
2. Trent, J.S., Scheinbeim, J.I. and Couchman, P.R. (1983), Macromols. 16, 589.
3. Parent, R.R. and Thompson, E.V. (1978) J. Polym. Sci.: Polym. Phys. *16*, 829.
4. Buchnall, C. and Partridge, I. (1983) Polymer *24*, 639.
5. Mendelson, R.A. (1985), J. Polym. Sci.: Polym. Phys. 23, 1975.
6. Butta, E., Levita, G., Marchetti, A.and Lazzeri, A. (1986) Polym. Eng. Sci. *26*, 63.
7. Venugopal, G. and Krause, S. (1992) Macromols. *25*, 4626.
8. Hashimoto, T., Shibayama, M. and Kawai, H. (1980) Macromols. *13*, 1237.
9. Hashimoto, H., Fujimura, M., Hashimoto, T. and Kawai, H. (1981) Macromols. *14*, 844.
10. Hashimoto, T.,Tsukahara, Y., Tachi, K. and Kawai, K. (1983) Macromols. *16*, 648.
11. Alward, B., Kinning, D.J., Thomas, E.L. and Fetters, L.J. (1986) Macromols. *19*, 215.
12. Meier, D.J. (1969) J. Polym. Sci. C *26*, 81.
13. Li, W. and Huang, B. (1992) J. Polym. Sci.: Polym. Phys. *30*, 727.
14. Mogi, Y., Hiroyuki, K., Kaneko, Y., Mori, K., Matsushita, Y. and Noda, I. (1992) Macromols. *25*, 5408.
15. Hashimoto, T., Yamasaki, K., Koizumi, S. and Hasegawa, H. (1993) Macromols. *26*, 2895.
16. Chen, X., Gardella, Jr., J.A. and Kumler, P.L. (1992) Macromols. *25*, 6631.
17. Sakurai, S., Momii, T., Taie, K., Shibayama, M. and Nomura, S. (1993) Macromols. *26*, 485.
18. Collin, B., Chatenay, D., Coulon, G., Ausserre, D. and Gallot, Y. (1992) Macromols. *25*, 1621.
19. Annis, B.K., Schwark, D.W., Reffner, J.R., Thomas, E.L. and Wunderlich, B. (1992) Makromol. Chem. *193*, 2589.
20. Arai, K., Kotaka, T., Kotano, Y. and Yoshimura, K. (1980) Macromols. *13*, 1670.
21. Kudose, I. and Kotaka, T. (1984) Macromols. *17*, 2325.
22. Gido, S.P., Schwark, D.W.S. and Thomas, E.L. (1993) Macromols. *26*, 2636.
23. Auschra, C. and Stadler, R. (1993) Macromols. *26*, 2171.
24. Beckmann, J., Auschra, C. and Stadler, R. (1993), Makromol. Chem. Rapid Commun. *15*, 67.
25. Mogi, Y., Mori, K., Matsushita, Y. and Noda, I. (1992) Macromols. *25*, 5412.
26. Gebizlioglu, O., Argon, A. and Cohen, R. (1985) Polymer *26*, 529.
27. Spontak, R.J., Smith, S.S. and Ashraf, A. (1993) Polym. Comm. *34*, 2233.
28. Winey, K.I., Thomas, E.L. and Fetters, L.J. (1992) Macromols. *25*, 422.
29. Hashimoto, T., Koizumi, S., Hasezawa, H., Izumitani, T. and Hyde, S.T. (1992) Macromols. *25*, 1433.

30. Hudson, S.D., Davis, D.D. and Lovinger, A.J. (1992) Macromols. *25*, 1759.
31. Li, W., Yan, R. and Jiang, B. (1992) Polymer *33*, 889.
32. Martuscelli, E., Silvestri, C., Addonizio, M.L. and Amelino, L. (1986) Makromol. Chem.*187*, 1557.
33. Vaughan, A.S. (1992) Polymer *33*, 2513.
34. Lovinger, A.J. and Williams, M.L. (1980) J. Appl. Polym. Sci. *25*, 1703.
35. Sundararajan, P.R., Gerroir, P.J., Bluhn, T.L. and Piche, Y. (1991) Polym. Bull. *27*, 345.
36. Organ, S.J. and Barham, P.J. (1993) Polymer *34*, 459.
37. Conde Brana, M.T. and Gedde, U.W. (1992) Polymer *33*, 3123.
38. Lotz, B., Kovacs, A.J., Bassett, G.A. and Keller, A. (1966) Kolloid Z. u. Z. Polym. *209*, 115.
39. Lovinger, A.J., Han, B.J., Padden, Jr., F.J. and Mirau, P.A. (1993) J. Polym. Sci.: Polym. Phys. *31*, 115.
40. Tsitsilianis, C., Strarkos, G., Dondos, A., Lutz, P. and Rempp, P (1992) Polymer *33*, 3369.
41. Serpico, J.M., Wnek, G.E., Krause, S., Smith, T.W., Luca, D.J. and Van Laeken, A. (1992) Macromols. *25*, 6373.
42. Janssen, K., Van Beylen, M., Samyn, C. Scherrenberg, R. and Raynaers, H. (1990) Makromol. Chem. *191*, 2777.
43. Stellman, J.M. and Woodward, A.E. (1969) J. Polym. Sci. Part B *7*, 755.
44. Odian, G. (1991) *Principles of Polymerization.* 3rd Ed. New York: Wiley-Interscience.
45. Zemel, I.S., Corrigan, J.P. and Woodward, A.E. (1989) J. Polym. Sci.: Polym. Phys. *27*, 2479.
46. Corrigan, J.D., Zemel, I.S. and Woodward, A.E. (1989) J. Polym. Sci.: Polym.Phys. *27*, 1135.
47. Hu, X., Stanford, J.L., Day, R.J. and Young, R.J. (1992) Macromols. *25*, 672.
48. Lindberg, K.A.H. and Bertikson, H.E. (1991) J. Mat. Sci. *26*, 4383.
49. Zhu, L. and Wegner, G. (1981) Makromol. Chem. *182*, 3625.
50. Briber, R.M. and Thomas, E.L. (1985) Polymer *26*, 8; see also (1986) Polymer *27*, 66.
51. Samuels, S.L. and Wilkes, G.L. (1973) J. Polym. Sci., Polym. Symp. *43*, 149.
52. Wegner, G., Zhu, L. and Lieser, G. (1981) Makromol. Chem. Rapid Comm. *182*, 231.
53. Allcock, H.R. and Lampe, F.W. (1990) *Contemporary Polymer Chemistry*, 2nd Ed. Englewood Cliffs, NJ: Prentice-Hall.
54. Friedrich, K., Walter, R., Voss, H. and Karger-Kocsis, J. (1986) Composites *17*, 205.
55. Kosfield, R. and Uhlenbroich, T. (1988) in *Atlas of Polymer Morphology*. Munich: Hanser Publishers, pp. 314, 315, 359, 361.
56. Sawyer, L.C. and Grubb, D.T. (1987) *Polymer Microscopy*. London: Chapman and Hall.
57. Schulte, K., Kutter, S. and Friedrich, F. (1988) DFVLR Report 1986 in *Atlas of Polymer Morphology*. Munich: Hanser Publishers, pp. 312, 314.

6 Morphologies Associated with Processing

Noncrystallizable polymers consist of separate but overlapping chains in random coil conformations or exist as crosslinked networks; these can be rubbery or rigid glassy matrices. As discussed in Chapter 3, quiescently melt-crystallized polymers usually yield lamellar structures such as hedrites and spherulites. If deformation is applied during the crystallization of thin films, a fibrous or epitaxial morphology is adopted. Composites and various block copolymers and blends form multiphase systems. The processing of relatively thick samples to form useful objects can involve both quiescent and deformational crystallization, leading to a mixture of morphologies, as described in this chapter.

Polymer objects can be fabricated in various shapes and sizes, as film, tubes, fabrics, and shaped objects, by extrusion or molding. The morphologies resulting from the use of particular processing methods and conditions are discussed below.

6.1   Skin/Core and Related Morphologies

The processing of a linear-chain polymer that is glassy or partially crystalline at use temperatures requires heating to soften pieces or sheets of the material being fabricated as well as the application of pressure to facilitate flow and/or packing of the melt in the extruder or the mold. Either as or after the object is formed, it is subjected to cooling and a reduction in pressure. In thick samples, a temperature gradient can exist during the cooling phase, causing the outer portions to cool more rapidly than the innermost parts. For extruded and injection-molded materials, such cooling can lead to different morphologies at the sample surface (the skin) and in the interior (the core), or to a trilayer structure containing a skin, an intermediate region and a core [1].

Extrusion through an opening or the injection of a polymer melt into a mold by the application of pressure causes chains to assume elongated conformations in the direction of the applied stress. A higher degree of orientation is usually achieved with injection molding due to the forces involved. If the surface of the mold is cold and the sample is thin, solidification will occur quickly, freezing in some or all of the chain orientation throughout the sample. If the surface of the mold is cold but the sample is thick, cooling of the polymer in the interior is slower than near the surface, and morphological differences will be observed due to a return to less elongated conformations of the interior chains prior to solidification. The center portion of thick samples (the core) will contain nondeformed spherulites and the surface (the skin) will consist of highly oriented chains. For some samples, the spherulites in a region between the skin and the core have conical shapes due to thermal gradients that occur during their formation [1]. A drawing depicting a skin/intermediate/core morphology is given in Fig. 6.1. The skin/intermediate/core morphology has been observed for polyethylene, polyoxymethylene and isotactic polypropylene [1]. Injection molded and extruded samples of some crystallizable polymers, such as polyamide 66 and poly(butylene terephthalate), have a nonoriented noncrystalline skin and a spherulitic core [1]. These two polymers contain polar groups which result in high glass transitions. Upon cooling, the outer portions of the sample quench to a glass before crystallization can occur there.

Lines and boundaries can occur between layers of material in molded objects due to rapid solidification of the flowing material. Flow lines have been observed in both the skin and the core parts of molded samples [1].

Spherulitic crystallization in thin polymer melts subjected to a thermal gradient has been investigated using optical microscopy with crossed polaroids

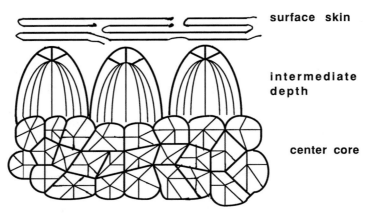

Figure 6.1   Skin/core/intermediate morphology from injection molding.

[2, 3]. Crystallization under an imposed temperature gradient yields extended lamellar growth in the direction of the zone motion and a distortion of the symmetry to give conically-shaped spherulites, as observed for the middle layer of the trilayer morphology described above (see Fig. 6.1).

6.2   Epitaxial Lamellar Morphology

Application of a uniaxial tensile stress during bulk crystallization of a thin film can give lamellas with the crystallographic c axis along the deformation direction [4, 5]. TEM investigation of uniaxially melt drawn and annealed polyethylene films shows that adjacent lamellas have many chains in common. The optimum molecular weight (M_w) for melt drawing polyethylene was found to be $4–9 \times 10^5$ at 130–160 °C at a moderately high strain rate [6].

Extrusion of a flexible chain polymer, such as polyethylene, through a small orifice, such as present in a capillary rheometer, yields tapered interlocking lamellas epitaxially crystallized on extended chains [7]. The lamellas are arranged perpendicular to the extrusion direction, while the extended chains and the crystal stems in the lamellas are oriented along the extrusion direction. The lamellar tapering taking place during crystallization is attributed to a temperature gradient existing in the extruder. Flow of a semi-rigid liquid crystalline copolymer of 4-hydroxybenzoate and 6-hydroxynapthenate through a capillary leads to chain extension [8]; however, better orientation is obtained by melt drawing.

6.3   Torsional Flow Morphology

Crystallization of polyethylene under torsional flow conditions along with the application of some compression gives samples having circular chain orientation, as shown by scanning electron microscopy (SEM) [9].

6.4   Blow Molding Morphology

Polyethylene film formed by blow molding with extrusion at a high draw ratio was found using SEM to consist of separated piles of 3–10 lamellar sheets with each pile functioning as a single entity [10]. The lamellas are stacked normal to the extrusion direction and adjacent piles are connected by fibrils in the extrusion direction. The blow molding process imparts a biaxial orientation to the film due to sequential extension in two perpendicular directions.

6.5   Blend and Composite Morphologies

The morphology of ternary blends of polycarbonate (the predominant component) with a more brittle polymer, such as atactic poly(methyl methacrylate) (PMMA), atactic polystyrene, or a styrene/acrylonitrile statistical copolymer, and a smaller amount of a compatibilizer, a material composed of butadiene/styrene rubber particles surrounded by a grafted shell of poly(methyl methacrylate) have been investigated [11]. Three statistical copolymers of styrene/acrylonitrile were studied: SAN17.5, SAN25 and SAN34, containing 17.5, 25 and 34% acrylonitrile, respectively. The distribution of compatibilizer particles, in the polycarbonate phase, in the brittle polymer phase and at the interface between the two for a 60/30/10 mixture of polycarbonate (PC), to brittle polymer (BP), to compatibilizer (MBS), was found to depend on the brittle polymer used and the mixing process, carried out in an extruder. Mixing was accomplished in four different ways, simultaneous mixing of all three components or mixing of two components followed by addition of the third. When BP is atactic PMMA, the MBS is located entirely in the PMMA phase due to the chemical similarities of the MBS outer shell and the PMMA. For the other brittle polymers, simultaneous mixing yields blends for which the MBS particles are located mainly at the interfaces when BP is atactic polystyrene, at the interfaces and in both phases when BP is SAN25, completely in the BP phase for SAN17.5, and completely in the PC phase for SAN34. Changes in the size and in some cases changes in shape of the particles present in PC/BP binary blends were observed upon simultaneous mixing with MBS. For polystyrene and for SAN34, spherical particles were present in both the binary and tertiary blends, but with larger diameters in the tertiary blends. For SAN17.5 and SAN25, the binary blend contained spheres and irregularly shaped particles while the tertiary blend showed enlarged irregularly shaped particles. Sequential mixing of either PC or BP with MBS followed by addition of the other component yielded larger more irregular domains, when BP was atactic polystyrene. The distributions obtained for the various mixing sequences, when BP is SAN25, are given in Table 8.1. It was found that the location of the MBS particles in the blends could be approximately predicted from the values of the pair interfacial energies and pair

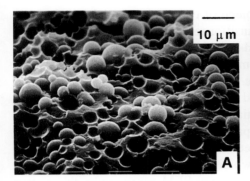
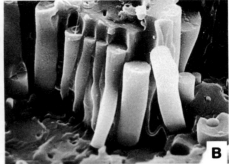

10 µm

A

B

Figure 6.2   Scanning electron micrograph of cryogenic fracture surfaces from blends of 40/60 w/w polyamide 6 with polyethylene/vinyl acetate copolymer extruded at 220 °C with different residence times: A. 3 min., center part; B. 3 min, border region. Courtesy of E. Martuscelli [Reprinted from: D'Orazio, L., Mancarella, C., Martuscelli, E., Casale, A., Filippi, A. and Speroni, F. (1986) J. Mat. Sci. *21*, 989].

miscibilities of the components. As discussed in Chapter 8, the impact strength is related to the MBS particle distribution.

The morphology that results on the extrusion of a blend also depends on the composition and the processing parameters, such as the flow rate, the temperature and the relative viscosities of the components. If the viscosities vary, the component with the lower viscosity, such as found for liquid crystalline polymers, will generally be highly deformable. Blends of a liquid crystalline copolymer, 4-hydroxybenzoate and 6-hydroxynapthenate, and an essentially noncrystallizing homopolymer, polycarbonate, with a high T_g, give either spherical inclusions or fibers of the liquid crystalline polymer in extrudates [12]. Fiber formation increased with increasing flow rate and at low copolymer amounts. Blends of a modified poly(phenylene oxide), a polymer with a high T_g, with a liquid crystalline polymer leads to the production of fine fibrils of the latter [13].

When blends of two crystallizable polymers, poly(ε-caprolactam) and an ethylene/vinyl acetate copolymer, were extruded using a capillary rheometer, the morphology depended on the composition, the processing temperature, and the time in the rheometer [14]. Irregularly shaped rods of one component near the outer portions of the extruded filament were apparent using SEM; in the interior of this extruded filament spherically shaped particles of this component were found, as seen in Fig. 6.2. For extruded blends of polyethylene and poly(ε-caprolactam) the particles in the dispersed phase elongate and then become fibrous [15].

The injection molding of a 90/10 blend of two noncrystallizing polymers, one a homopolymer with a high glass transition temperature and the other a block copolymer, was found to give a ball-and-string structure in the outer parts of the sample after removal of the homopolymer, the principle component [16]. The glassy homopolymer is polycarbonate and the block copolymer contains a polystyrene/polyacrylonitrile statistical copolymer as the rigid portion and polybutadiene, as the rubbery part. Partial degradation of the block copolymer, occurring during molding gave a rigid part (the strings) and a rubbery part (the balls) when the extruded material cooled. The morphology at the center of the molded plaque contained particles of two different diameters derived from the block copolymer dispersed in the homopolymer. SEM micrographs, taken after removal of the homopolymer matrix on fracture surfaces obtained at various positions in the sample, showing these two morphologies, are given in Fig. 6.3.

Injection molding of rigid fiber containing composites can lead to alignment of the fibers parallel to the surface of the sample in the skin with no alignment in the sample core [1].

For injection molded polypropylene/mica composites, the mica flakes in the skin portion orient parallel to the mold surface, while the orientation in the interior is more perpendicular [17].

6.6   Processing from Solution

Processing can involve the recovery of a crystallizing polymer from solution by solvent removal. The resulting morphology will depend on the polymer chain conformation in the solution. If the polymer molecules are in a random coil conformation and the solution is isotropic, crystallization occurring during solvent evaporation usually leads to spherulite growth. If a high degree of chain orientation occurs in the solution, as is possible in some lyotropic liquid crystalline mesophases of semi-rigid chain polymers such as poly(p-phenylene terephthalamide), then solvent removal is expected to yield a highly ordered sample [18]. However, as discussed in the following section, relaxation effects can occur.

6.7   Fiber Banding

The appearance of periodic bands, about 0.5–10 µm wide, oriented perpendicular to the fiber direction, has been reported for samples obtained from lyotropic

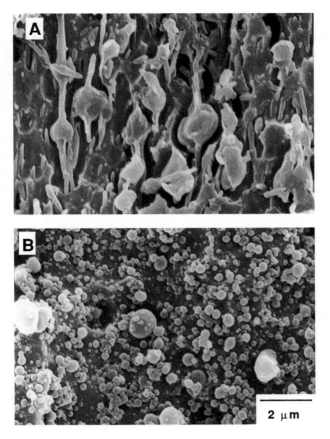

Figure 6.3   Scanning electron micrographs of injection-molded 90/10 blend of polycarbonte with acrylonitrile-butadiene-styrene copolymer after brittle fracture and etching with NaOH: A. bead-string morphology 1.5 mm from sample center, B. 0.13 mm from center. Courtesy of A. Hiltner [Reprinted from: Lee, M.P., Hiltner, A. and Baer, E. (1992) Polymer *33*, 675].

and thermotropic liquid crystalline mesophases subjected to shearing or a flow field [19–25]. This texture is commonly observed for polymers with main-chain mesogens but has also been reported for polymers with rigidly attached side-chains [25]. Banding is observed to evolve following cessation of the flow [21, 22]. In one lyotropic system, polybenzothiazole in 9.8% polyphosphoric acid solution, alignment using a flow field also resulted in orientation of the solvent molecules [21]; upon removal of the field and evaporation of the solvent, three stages of relaxation of the orientation were observed. The first of these stages was solvent disorientation; this was followed by a period of slow stress relaxation, mesogen disorientation with a periodic variation in the director and band formation; finally, a very slow process involving transition from banding to a polydomain situation occurs. The bands represent a periodic fluctuation of the molecular orientation (and the director) around the shear direction. This fluctuation is due to an alternating tilt of the optic axis by equal and opposite amounts relative to the shear direction and is caused by strains due to contraction after flow is halted [26].

The bands appearing in a nematic lyotropic mesophase of a polydiacetylene consisted of a herringbone arrangement of lamellas when the solvent was completely removed [27]. When films were left in the lyotropic state for a week, disclination pairs were observed due to relaxation of the original orientation.

For some fibers obtained from liquid crystalline mesophases, the banding observed in oriented systems disappeared upon annealing [1]. Shearing during crystallization of a thermotropic liquid crystalline polyester yielded fibrils in a zig-zag array [28].

6.8   Relaxed Spherulite Growth

Compression molding, which involves melting of the polymer in a mold followed by the application of an equal pressure over all portions of the sample, should not lead to the adaptation of extended chain conformations during the molding process. Polymers crystallizing under these conditions should have morphologies similar to those freely crystallized from the melt. In fact, this is what is observed for a compression-molded polyethylene copolymer crystallizing as banded spherulites [29].

6.9   Crosslinked Networks

Rigid amorphous crosslinked (thermosetting) polymer networks can be prepared from chemically active low molecular weight precursors by compression molding and by bag molding, in which the crosslinked resin is reinforced. Pressure is applied in compression and bag molding to pack the sample and thereby prevent void formation. Thermoplastic polymers containing reactive groups can be converted into crosslinked networks by heating in the presence of a crosslinking agent.

Homogenization of a thermosetting epoxy resin/rubber blend was found to occur when a shear force is applied to the melt [30]. When the blend is lightly sheared, streak patterns appear in the shear direction; when strongly sheared, particularly just prior to gelation, the system appears homogeneous.

6.10   Voids in Polymer Solids

Various sized voids are observed in extruded and molded samples due mainly to incomplete packing of polymer fluid [1]. Voids near the surface of extruded samples are found to be elongated in the direction of the applied force.

The spherulites or hedrites, formed under quiescent melt-crystallization conditions in thin films and present near the surface, do not completely fill the space available. Small spaces between and within the hedrites and spherulites, caused by material depletion during the crystallization process, are observed using SEM (see Chapter 3).

6.11   Environmental Effects

Various morphological effects on polymeric systems caused by temperature changes, stress application, electrical and magnetic fields, solvents, reactive chemicals and ion etching have been discussed in previous chapters. In this section, some further examples of environmental effects will be discussed.

Oxidative degradation can take place when a predominantly hydrocarbon polymer is heated in air. Extruded polyethylene pipe containing carbon black, when cooled in air from >200 °C, has a nonspherulitic oxidized layer 0–20 μm thick, an intermediate region and an inner part made up of large spherulites [31].

Poly(vinylidene fluoride) film crystallized at 160 °C for 15 days developed discolored areas, apparently due to dehydrofluorination [32].

Isotactic polypropylene develops an embrittled surface layer when subjected to ultraviolet, UV, irradiation in air [33]. When compared to injection-molded samples, compression-molded material exhibited a deeper embrittled layer and a larger number of carbonyl groups, caused by reactions with oxygen. The compression-molded samples were formed at a higher temperature and exposed to oxygen longer than injection-molded ones; also, the injection-molded samples have a greater degree of orientation, enhancing the crystallinity.

References

1. Sawyer, L.C. and. Grubb, D.T. (1987) *Polymer Microscopy*. London: Chapman and Hall.
2. Lovinger,A.J. and Gryte, C.C. (1976) Macromols. *9*, 247.
3. Lovinger, A.J., Chua, J.O. and Gryte, C.C. (1977) J. Polym. Sci.: Polym. Phys. *15*, 641.
4. Gohil, R.M. and Petermann, J. (1982) J. Colloid Polym. Sci. *260*, 312.
5. Adams, W.W., Yang, D. and Thomas, E.L. (1986) J. Mat. Sci. *21*, 2239.
6. Bashir, Z. and Keller, A. (1989) Colloid and Polym. Sci. *267*, 116.
7. Bashir, Z., Hill, M.J. and Keller, A. (1986) J. Mat. Sci. Lett. *5*, 876.
8. Kohli, A., Chung, N. and Weiss, R.A. (1989) Polym. Eng. and Sci. *29*, 573.
9. Kanamoto, T. and Zachariades, A.E. (1985) Polym. Eng. Sci. *25*, 494.
10. Tagawa, T. and Ogura, K. (1980) J. Polym.Sci.: Polym. Phys. Ed. *18*, 971.
11. Cheng, T.W., Keshkula, H. and Paul, D.R. (1992) Polymer *33*, 1606.
12. Isayev, A.I. and Modic, M. (1987) Polymer Composites *8*, 158.
13. Ogata, N., Yu, H.V., Ogihara, T., Yoshida, K., Kondou, Y., Hayashi, K. and Yoshida N. (1993) J. Mat. Sci. *28*, 3228.
14. D'Orazio, L., Mancarella, C. , Martuscelli, E., Casale, A. Filippi, A. and Speroni, F. (1986) J. Mat. Sci. *21*, 989.
15. Gonzalez-Nunez, R., Favis B.D., Carreau, P.J. and Lavallee C. (1993) Polym. Eng. Sci. *33*, 851.
16. Lee, M.P., Hiltner, A. and Baer, E. (1992) Polymer *33*, 675.
17. Xavier, S.X., Schultz, J.M. and Friedrich, K. (1990) J. Mat. Sci. *25*, 2411.
18. Dobb, M.G., Johnson, D.J. and Saville, B.P. (1977) J. Polym. Sci.: Polym. Symp. *58*, 237.
19. Donald, A.M. (1988) in *Atlas of Polymer Morphology*. Munich: Hanser Publishers, pp. 269, 290.
20. Donald, A.M., Viney, C. and Windle, A.H. (1983) Polymer *24*, 155.
21. Odell, J.A., Ungar, G. and Feijoo, J.L. (1993) J. Polym. Sci.: Polym. Phys. Ed. *31*, 141.
22. Picken, S.J., Moldenaers, P, Berghmans, S. and Mervis, J. (1992) Macromols. *25*, 4759.
23. Sawyer, L.C. and Jaffe, M. (1986) J. Mat. Sci. *21*, 1897.
24. Shimamura, K. (1983) Makromol. Chem. Rapid Comm. *4*, 107.
25. Xu, G., Wu, W., Shen, D., Hou, J., Zhang, S., Xu, M. and Zhoum Q. (1993) Polymer *34*, 1818.
26. Viney, C., Donald, A.M. and Windle, A.H. (1985) Polymer *26*, 870.
27. Wang, W., Lieser, G. and Wegner, G. (1993) Makromol. Chem. *194*, 1289.
28. Chen, S.S., Hu, S.R. and Xu, M. (1988), in *Atlas of Polymer Morphology*. Munich: Hanser Publishers, pp. 269, 290.
29. Bubeck, R.A. and Baker, H.M. (1982) Polymer *23*, 1680.
30. van Dijk, M.A., Eleveld, M.B. and van Veelan, A. (1992) Macromols. *25*, 2274.
31. Terselius, B., Gedde, U.W. and Jansson, J.-F. (1982) Polym. Eng. Sci. *22*, 422.
32. Lovinger, A.J. and Freed, D.J. (1980) Macromols. *13*, 989.
33. Schoolenberg, G.E. and Vink, P. (1991) Polymer *32*, 432.

7 Morphologies Associated with Deformation and Fracture

7.1 Deformation Effects

Upon tensile stress application isotropic rubbers will easily deform, giving elongated chain conformations; a rubber specimen will recover its original dimensions and conformational distribution on stress release, if the system is a crosslinked one. Glassy polymer materials form crazes, shear bands, or a combination thereof, and certain microcrystalline polymers will yield and undergo relatively large amounts of plastic deformation that remains after stress release.

7.1.1 Elastic Deformation

Materials containing both rubbery and glassy components can be elastically deformed, and when the stress is released they return to their original dimensions. This type of behavior is exhibited by a triblock copolymer containing 27% polystyrene/30% polybutadiene/43% poly(4-vinyl pyridine), having spheres of poly(4-vinyl pyridine), a glassy constituent, surrounded by spherical shells of polybutadiene rubber in a continuous glassy polystyrene matrix when cast from chloroform solution [1]. The application of a fivefold uniaxial deformation leads to elongated regions of all three components in the stress direction; upon release of the stress the sample returns to the original morphology.

When a polystyrene/polybutadiene/polystyrene block copolymer is subjected to a shear stress applied at a right angle to the direction of the polybutadiene cylinders, the cylinders were rotated by 90° into the shear direction; maximum alignment occurs at intermediate strains ($\gamma = 7.8$) with deterioration of the order occurring at large strains ($\gamma = 12$) [2].

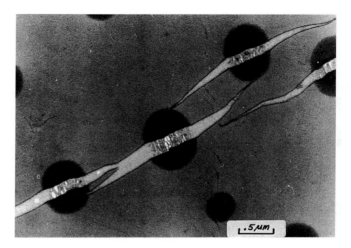

Figure 7.1   Transmission electron micrograph of a stressed film of a 96/4 poly(methyl methacrylate)/polystyrene blend cast from toluene and treated with RuO_4. Courtesy of J.I. Scheinbeim. [Reprinted with permission from: Trent, J.S., Scheinbeim, J.I. and Couchman, P.R. (1983) Macromols. *16*, 589. Copyright © 1983, American Chemical Society.

7.1.2   Craze Development

Crazes are networks of voids and oriented fibrillar elements involving localized plastic deformation occurring in glassy polymer solids [3]; they are usually stress related but can also be induced by reactive environments. Crazes are believed to be formed at surface flaws or in the sample interior at the surface of impurity particles and particles of incompatible components purposely added. Crazes are visually observable in transparent sheets of glassy polymers, such as polystyrene, as a network of white lines after a stress has been applied and removed. At higher magnifications, the parallel fibrillar nature of a craze is apparent [4]. Cross ties on craze fibrils have been observed [5]. A craze region usually has a width of about 1 μm or less along the fibril direction. The craze length is usually at least five to six times larger than the craze width. An example of craze formation in a film prepared from a blend of 96% atactic poly(methyl methacrylate) and 4% atactic polystyrene after subjecting it to a tensile stress and treating with RuO_4 is seen in the transmission electron microscopy (TEM) micrograph in Fig. 7.1 [6]; each craze shown in this photo is associated with a polystyrene particle, and the four crazes are oriented in the same direction.

   Tensile deformation of an injection molded blend of 25% ABS rubber/25% styrene-maleic anhydride-methyl methacrylate copolymer/50% polycarbonate leads to failure at a strain of 133% [7]. (ABS is a block copolymer of poly(acrylonitrile/styrene) with polybutadiene.) The ductile polycarbonate-rich phase stops the abundant crazes observed; ABS rubber particle deformation and matrix phase striations occur. Crazes in a blend of polystyrene and a polystyrene/polybutadiene star block copolymer initiated by tensile deformation were found to take various pathways [8]; some passed through the block copolymer particles, some were diverted and some were highly wavy.

7.1.3   Shear Band Formation

Another stress related effect that can develop in a glassy polymer is the formation of shear bands [9]. Shear bands involve the lateral movement of sections of material without voiding and are formed in glassy polymers under compressive stress when a shear force is applied to the specimen. Exposure to methanol liquid can induce crazes on the tension side of polystyrene blocks containing coarse shear bands caused by previous compression [10]. Shear banding and crazing can occur together under tensile loading in more ductile glassy polymers, such as polycarbonate [11].

7.1.4   Crystallized Systems

7.1.4.1   Uniaxial Tensile Stress

The effects of tensile stress on single lamellas, that have been deposited on a polymer film which is subsequently drawn, have been investigated for various polymers [12]. Lamellar thinning in the stress direction and the appearance of a void network in the thinned regions was observed at low stress amounts (10–40% deformation) in square poly(4-methyl pentene-1) lamellas adhering to a poly(ethylene terephthalate) film [13]; at higher deformations (25–52%), fibrils orient approximately in the deformation direction, shear bands occur at various angles to this, and buckling appears, more or less at right angles, as shown in Fig. 7.2. For lathe-shaped polypropylene lamellas, deformation yielded the above effects along with kinks or twin bands at lower deformations [14].

Upon deformation of polyethylene spherulites, obtained from concentrated solution and soaked in oil, both localized yielding (inhomogeneous) and elongation and yielding of all parts of the spherulite (homogeneous) were observed, with the two types usually occurring together [15]. Both inter- and intralamellar yielding occurred when the deformation was primarily inhomogeneous.

Figure 7.2   Transmission electron micrograph of isotactic poly(4-methyl pentene-1) lamella crystallized from 0.024 wt % toluene solution at 78 °C, cooled to 25 °C and deposited on poly)ethylene terephthalate) film that is subsequently drawn 52%. [Reprinted from: Woodward, A.E. and Morrow, D.R. (1969) J. Polym. Sci. A-2 7, 1651. Copyright © 1969, John Wiley and Sons, Inc.]

When a uniaxial tensile stress is applied to a strip of a quiescently melt-crystallized polymer sample, yielding can occur and a permanent deformation ensue. This is know as cold drawing, and some of the changes occurring can be easily observed by eye [16]. The deformation occurs in a central region, the necking region, with large portions of the sample on either end of the strip remaining unchanged. Spherulites present in the necking region are plastically deformed, first becoming elongated, then breaking up and rearranging to form a fibrillar mass. Voids appear, causing stress whitening in the necking region. Drawing of melt-crystallized polymer samples having entangled chains is limited to draw ratios of about 8.

The effects of drawing (draw ratio of about 4) on individual lamellas in banded polyethylene spherulites at 80 °C was found to be largest on those at the center of a spherulite [17]. Poly(butylene terephthalate) drawn three-and four-fold show lamellas perpendicular to and fibers in the draw direction [18]. The best orientation was obtained at a temperature between T_g and T_m.

The effects of tensile deformation on polyethylene, crystallized from the melt at 120 °C while subjected to a longitudinal flow field and then annealed at 128 °C, were studied for strains up to 350% [19]. Deformation along the chain direction causes the lamellas to separate and partly unfold forming fibrils parallel to the deformation direction. The fibrillar morphology resulting from the stress contains oriented crystals, crystal blocks and recrystallized lamellar chains.

Ultradrawn fibers, tapes, and films, in which there are few, if any, chain folds, can be prepared from flexible chain polymers starting with the appropriate sample. Polyethylene, as polymerized at 50 °C and drawn at 130 °C to a draw ratio of 60, resulted in fibers with a double orientation of the [c] crystallographic axis along the draw direction and the [a] crystallographic axis approximately perpendicular to the draw direction [20] . Solution cast polyethylene gel films drawn at 130 °C sustained draw ratios of 130 and gave irregular corrugations perpendicular to the draw direction and double crystallographic orientation [21, 22]. When these samples are melted, allowed to crystallize and then cold drawn, only a draw ratio of 8 can be sustained prior to fracture. The resulting fibers contain crystallites with chain folds.

When a narrow zone of a sample of a crystallized polymer, such as polyethylene, is heated to temperatures six to eight degrees above T_m and drawn, extensions up to 150 times are sustained at low stress amounts [23]. At low draw ratios, epitaxial growths of lamellas on extended chains are observed; as the draw ratio is increased, the morphology changes to one consisting of smooth fibrils. During zone drawing above T_m, transformation of the spherulitic lamellar structure to a fibrillar structure takes place in the upper shoulder of the neck, followed by flow into the heating zone where it is homogeneously drawn; upon leaving this zone, it quickly cools and crystallizes. Zone drawing with the temperature below T_m leads to a fibrillar structure with a minor amount of interfibrillar transverse lamellas being formed.

Cross sections of poly(ethylene terephthalate) fibers prepared by high-speed spinning are about one-third crystalline and contain fibrils separated by amorphous domains [24].

Polyethylene film subjected to drawing, rolling and annealing contains tilted lamellar stacks with the chain axis parallel to the draw direction [25].

7.1.4.2   Biaxial Tensile Stress

Biaxial stress applied to a film prepared by extrusion of a concentrated solution of polyethylene in decalin followed by water quenching and drying yields a product with a high degree of porosity [26]. The porosity is related to the density by:

$$porosity = [(\rho_o - \rho) / \rho_o] \, 100$$

where ρ_o is the density of biaxial oriented nonporous material and ρ is the density of the porous product. The porosity rises to over 80% at λ (biaxial) = 3×3, then falls off but stays above 80% to λ (biaxial) = 8×8. The resulting product is a loose network of highly oriented fibrils. Nonporous samples with fibrils oriented in one particular direction can be prepared by annealing the film before the biaxial orientation step.

7.1.4.3   Compressive Stress

Application of compressive stress to crystallized polymer samples leads to effects that depend on the type of compression used and the sample morphology. The effects of plastic plane strain compression of spherulitic polyethylene to compression ratios up to 12 were investigated [27]. At a compression ratio (CR) of 1.8, coarse shear bands were observed at angles of $(\pm)45°$ with respect to the flow direction; at a CR of 2.5, a dense collection of fine shear bands between coarse shear bands was evident; above a CR of 2.5, tilting of the shear bands occurs. Rows of kinked lamellas are observed at CR = 1.8, and at CR's of 2.5 and 3.13, the lamellas generally orient parallel to the flow direction. At CR = 3.13, the thinner lamellas show large amounts of fragmentation; above 3.13, lamellar fragments rearrange into new lamellas oriented at an acute angle about the load direction. For melt-crystallized poly(ε-caprolactam), plane strain compression ratios greater than 1.6 cause intense shear to appear at $(\pm)45°$ with respect to the flow direction, most frequently at interspherulitic boundaries [28]. A change of crystal form in chains along the shear direction and lamellar deterioration were also observed. Uniaxial compression of polyethylene leads to crystal slip as the major deformation mechanism [29]. Interlamellar shear, but no lamellar fragmentation, was observed.

Compaction of polyethylene fibers at 138 °C causes them to pack together with an irregular coordination number from 4 to 7 and with misalignment in successive rows [30]. The deformation is by shear, and lamellas are evident between and within fibers with the same chain orientation as the fibers. Hot compaction of melt-spun polyethylene fibers at pressures of 100 psi, then of 3,000 psi, led to the melting of the outer parts of the fiber and recrystallization as lamellas [31]. A herring-bone stacking of lamellas occurs with the chain direction approximately the same for the lamellas and the fibers.

7.2   Fracture Surfaces

The fracture surface appearance depends on the chemical and physical nature of the sample and the type of deformation applied.

7.2.1   Glassy Polymers: Tensile or Compressive Stress

The fracture of a glassy polymer, caused by application of an increasing amount of tensile stress, can be preceded by craze formation, generated in the high stress fields ahead of crack tips. The fracture surface is primarily smooth but can have numerous steps, due to propagation of the stress on more than one plane [32]. A scanning electron microscopy (SEM) micrograph of a fracture surface for an extruded polystyrene rod subjected to tension is show in Fig. 7.3. Fracture can be preceded by shear bands during the application of a steadily increasing amount of compressive stress. Four regions have been identified on a shear fracture surface of polycarbonate: 1) initiation, 2) a cloudy, or *mist*, region, where shear failure occurs, 3) a mirror region with closed conic markings, and 4) an end band [33].

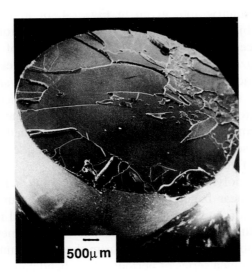

Figure 7.3   Scanning electron micrograph of a fracture surface for an extruded polystyrene rod subjected to tension. Courtesy of J.A. Sauer [as in: Sauer, J.A., McMaster, A.D. and Morrow, D.R. (1976) J. Macromol. Sci.-Phys. B *12*, 535].

The fracture surface of an oriented glassy polymer, poly(methyl methacry-late), drawn threefold above T_g, was found to contain dimples, fibrous sheets, and fibers, the latter indicating crack propagation along the draw direction [34].

Thick films of a polyphosphazene with semi-rigid (p-phenylphenoxy) side chains contain linked globules about 100 nm in diameter which form rods 15–40 nm wide when drawn four- to fivefold at 145 °C [35].

Environmental stress cracking occurred on the inside surface of medium density polyethylene pipe after immersion in a 1% aqueous surfactant solution at 50 °C and placement under constant load [36]. Cracks crossed spherulitic boundaries, and interlamellar separation was evident.

7.2.2   Glassy Polymers: Oscillatory Stress

Fracture can be induced by an oscillatory (time-dependent) stress using maximum stress levels below those necessary to break the sample by application of a tensile or compressive stress alone. The oscillatory stress can involve tension/less tension or tension/compression. Fracture induced in this manner is referred to as *fatigue fracture* and usually commences at the sample surface. For many polymers, the fatigue fracture surface contains a series of concentric crack growth bands surrounding the surface source [37]. An example of this feature is seen in the SEM micrograph, given in Fig. 7.4, of a fracture surface for polystyrene rod subjected to alternating tension/compression. The bands surrounding the source are caused by intermittent growth of the crack due to breakdown of a craze. More than one oscillation of the stress occurs between each jump of the crack. The *discontinuous crack growth bands* are followed by a region that may show radial tear lines, secondary fracture features and increasing surface roughness. The *fatigue lifetime*, the time necessary for the sample to fracture completely, decreases with an increase in the maximum stress and usually increases with increasing oscillation frequency, provided that excessive thermal effects do not occur.

For some glassy polymers exhibiting discontinuous crack growth due to an oscillatory stress, shear banding as well as crazing occurs [11]. Polymers showing this behavior, such as polycarbonate, are soft and more ductile than glassy polymers, such as polystyrene and poly(methyl methacrylate). Thin sections of these more ductile materials from the discontinuous crack region propagation stage show a series of periodic shear band pairs angled at approximately 45° to the main crack and to the stress direction; a craze is observed ahead of both the crack and the last pair of shear bands generated. The discontinuous fatigue crack propagation stage terminates when a microcrack develops in each

100 μm

Figure 7.4   Scanning electron micrograph of a fracture surface for an extruded polystyrene rod subjected to an alternating tension/compression stress of 17.2 MPa at a frequency of 2 Hz. Courtesy of J.A. Sauer [Reprinted from: Sauer, J.A. and Chen, C.C. (1983) Adv. Polm. Sci. *52/53*, 169].

of the final pair of shear bands. The shear band pair and the associated leading craze are called epsilon plastic zones. With the appearance of the two shear microcracks, the discontinuous stage of the fracture is over. Fracture advances in the direction of the microcracks with each stress cycle and this usually leads to shear-fatigue striations on the shear fracture surfaces.

7.2.3   Crystallized Polymers: Ultrasonic Vibration

When individual trans-1,4-polyisoprene hedrites, formed by crystallization from solution, are subjected to ultrasonic vibration in suspension, they split into two parts due to fracture across the middle along the short direction. All of the lamellas making up the stack are split in half [38]. Solution-grown spherulites of the same material do not fracture under an identical treatment. This suggests that the fracture is due to a high stress concentration at the center of the hedrite cluster where considerable lamellar curvature occurs.

7.2.4   Molded Samples of Crystallized Polymers: Tensile Stress

Fracture in molded crystalline polymers, such as polyethylene, subjected to a constant load can occur via a crack tip damage zone. Optical micrographs show many fibrils and interfibrillar cavities in this zone [39]. This is similar to craze-related fracture in glassy polymers. Pulled out fibrils have been observed on fracture surfaces of crystallized polymers [40]. Fracture in a crystalline polymer sample can occur both across and between spherulites [12]. For a polyethylene zone drawn at temperatures below T_m to give a fibrous structure, transverse cracks appear [23]; surface waves and kink bands near the fracture site have also been observed.

7.2.5   Molded Samples of Crystallized Polymers: Oscillatory Stress

Injection-molded samples of poly(hexamethylene adipamide) subjected to fatigue fracture exhibit multiple crazes around the crack tip [41]. Although, generally, these crazes pass through spherulites and cross spherulite boundaries, they occasionally change direction at the boundary.

7.2.6   Fibers

The types of fracture under tensile load observed for polymer fibers includes brittle fracture in elastomers, ductile fracture in some crystallized synthetic polymers, splitting along the fibril direction, and lateral failure [40]. Fractures that occur during the application of an oscillatory stress (fatigue fracture) are observed to commence at the fiber surface and to involve splitting along a pathway near the point of initiation. Compressive stress leads to kink band formation. Fracture of polyoxymethylene superdrawn fibers mainly affects the thin cross fibrils present within the thick fibrillar network [42].

The fracture surface of poly(p-phenylene terephthalamide) after being placed in refluxing hydrochloric acid for 8 hours showed perpendicular steps with heights of 200 nm and multiples thereof [43]. Drawn isotactic polypropylene monofilaments contained small regions that were resistant to reaction with chromic acid solution [44].

7.2.7   UV Irradiated Polymers

When either fibers or film of isotactic polypropylene, poly(ethylene terephthalate) or a polyamide are subjected to vacuum UV laser irradiation with a wavelength of 193, 248 or 157 nm (the latter for polypropylene), melting of the surface occurs and a ripple structure with raised squiggles a few microns high appears [45, 46]. When poly(p-phenylene terephthalamide) is subjected to UV irradiation at 310–800 nm for 7 days, spiral defect tracks and a ratchet-like surface are evident [47]. It was found that fracture occurred along the tracks.

7.2.8   Blends

The fracture surface for a blend containing inclusions of one material in a matrix of another is indicative of the adhesion at the surfaces between the two phases. The presence of holes in the matrix and separate particles of the dispersed material indicates poor adhesion, while fracture surfaces crossing both components shows good adhesion [16]. Blending of a compatible rubber with a crystalline/glassy polymer can change the mode of fracture from brittle to ductile for the latter and thereby change the fracture surface morphology from relatively smooth to rough [40].

7.2.9   Composites

For composites containing fibrous or particulate fillers, fracture usually leads to the exposure of numerous fiber surfaces. For some composites, these surfaces are smooth without any of the continuous polymer component adhering to the fiber surface, indicating poor adhesion (little attraction) between the two components. Fracture of the filler particles can also occur [40].

For glass fabric/epoxy laminates, failure involves in-plane cross-cracking with the exposure of voids [48].

References

1. Arai, K., Kotaka, T., Kitano, Y. and Yoshimura, K. (1980) Macromols. *13*, 1670.
2. Scott, D.B., Waddon, A.J, Lin, Y-G., Karasz, F.E. (1992) Macromols. *25*, 4175.
3. Kausch, H.H. (1983) *Advances in Polymer Science 52/53*.
4. Michler, G. H. (1985) Coll. Polym. Sci. *263*, 462.
5. Miller, P., Buckley, D.J. and Kramer, E.J. (1991) J. Mat. Sci. *26*, 4445.

6. Trent, J.S., Scheinbeim, J.I. and Couchman, P.R. (1983), Macromols. 16, 589.
7. Mendelson, R.A. (1985) J. Polym. Sci.: Polym. Phys. *23*, 1975.
8. Gebizlioglu, O., Argon, A. and Cohen, R. (1985) Polymer *26*, 519, 529.
9. Chau, C.C. and Li, J.C.M. (1981) J. Mat. Sci. *16*, 1858.
10. Chau, C.C. and Li, J.C.M. (1983) J. Mat. Sci. *18*, 3047.
11. Takemori, M.T. and Kambour, R.P. (1981) J. Mat. Sci. Lett. *16*, 1108 (1981).
12. Wunderlich, B. (1973) *Macromolecular Physics, Vol. 1*. New York: Academic Press.
13. Woodward, A.E. and Morrow, D.R. (1969) J. Polym. Sci. A2 *7*, 1651.
14. Cerra, P., Morrow, D.R. and Sauer, J.A. (1969) J. Macromol. Sci.: Phys. B3, *33*.
15. Hay,I.L. and Keller, A. (1965) Kolloid Z. u. Z. Polym. *204*, 43.
16. Woodward, A.E. (1988) *Atlas of Polymer Morphology*. Munich: Hanser Publishers.
17. Shimamura, K., Murakami, S. and Katayama, K. (1982) Makromol. Chem. Rapid Comm. *3*, 192.
18. Zhou, Z., Cackovic, H., Schultze, J.D. and Springer, J. (1993) Polymer *34*, 494.
19. Adams, W.W., Yang, D. and Thomas, E.L. (1986) J. Mat. Sci. *21*, 2239.
20. Smith, P., Chanzy, H.D. and Rotzinger, B.P. (1987) J. Mat. Sci. *22*, 523.
21. Smith, P., Lemstra, P.J., Pijpers, J.P.L. and Kiel, A.M. (1981) Colloid and Polym. Sci. *259*, 1070.
22. Smith, P, Boudet, A. and Chanzy, H. (1985) J. Mat. Sci. Lett. *4*, 13.
23. Hoff, M. and Pelzbauer, Z. (1991) Polymer *32*, 999.
24. Tomlin, D.W., Roland, C.M. and Slutsker, L.I. (1993) J. Polym. Sci.: Polym. Phys. *31*, 1331.
25. Grubb, D.T., Duglosz, J. and Keller, A. (1975) J. Mat. Sci. Lett. *10*, 1826.
26. Gerrits, N.S.J.A. and Lemstra, P.J. (1991) Polymer *32*, 1770.
27. Galeski, A., Bartczak, Z., Argon, A.S. and Cohen, R.E. (1992) Macromols. *25*, 5705.
28. Galeski, A., Argon, A.S. and Cohen, R.E. (1991) Macromol. *24*, 3953.
29. Bartczak, Z., Cohen, R.E. and Argon, A.S. (1992) Macromols. *25*, 4692.
30. Olley, R.H., Bassett, D.C., Hine, P.J. and Ward, I.M. (1993) J. Mat. Sci. *28*, 1107.
31. Hine, P.J., Ward, I.M., Olley, R.H. and Bassett, D.C. (1993) J. Mat. Sci. *28*, 316.
32. Sauer, J.A., McMaster, A.D. and Morrow, D.R. (1976) J. Macromol. Sci., Phys. *B12*, 535.
33. Agrawal, C.W., Hunter, K., Pearsall, G.W. and Henkens, R.W. (1992) J. Mat. Sci. *27*, 2606.
34. Lin, C.B., Hu, C.T. and Lee, S. (1993) Polym. Eng. Sci. *33*, 431.
35. Kojima, M. and Magill, J.H. (1992) Makromol. Chem. *193*, 379.
36. Lustiger, A. and Markham, R.L. (1983) Polymer *24*, 1647.
37. Sauer, J.A. and Chen, C.C. (1983) Adv. Polym. Sci. 52/53, 169.
38. Xu, J. and Woodward, A.E. (1986) Macromols. *19*, 1114.
39. Brown, N. and Bhattacharya, S.K. (1985) J. Mat. Sci. *20*, 4553.
40. Sawyer, L.C. and Grubb, D.T. (1987) *Polymer Microscopy*. London: Chapman and Hall.
41. Wyzgoski, M.G. and Novak, G.E. (1992) Poly. Eng. Sci. *32*, 1114.
42. Komatsu, T. (1993), J. Mat. Sci. 28, 3035.
43. Morgan, R.J., Pruneda, C.O., and Steele, W.J. (1983) J. Polym. Sci.: Polym. Phys. *21*, 1757.
44. Garton, A., Carlsson, D.J., Sturgeon, P.Z. and Wiles, D.M. (1977) J. Polym. Sci. Polym. Phys. *15*, 2017.
45. Kesting, W., Bahners, T. and Schollmeyer, E. (1993) J. Polym. Sci.: Polym. Phys. Ed. *31*, 887.
46. Wefers, L. and Schollmeyer, E. (1993) J. Polym. Sci.: Polym. Phys. Ed. *31*, 23.
47. Dobb, M.G.. Robson, R.M. and Roberts, A.H. (1993) J. Mat. Sci. *28*, 785.
48. Samajdar, S. (1991) J. Mat. Sci. *26*, 977.

8 Morphological Effects on Polymer Properties

The physical, mechanical, optical and electrical properties of polymer materials are affected to various degrees by the sample morphology. Physical properties, such as the density, the heat of fusion, as well as spectroscopic and x-ray diffraction intensities, for a particular polymer system depend on the level of crystallinity, which is a function of the crystallization conditions. Mechanical properties, such as the tensile strength, the tensile elongation and the impact strength, and optical properties, such as birefringence and transmittance, for a particular polymer vary with the morphology and, to a lesser extent, with the crystallinity. The physical, mechanical, optical and electrical properties can be affected by the type of fabrication method used due to orientation effects, the introduction of voids, and the presence of thermal gradients during cooling.

8.1 Crystallinity and Related Properties

The crystallinity of a flexible chain homopolymer, such as polyethylene, increases with the preparation method in approximately the following order:

1. Quenched quiescent melt crystallization (high degree of supercooling),
2. quiescent melt crystallization at small degrees of supercooling,
3. solution crystallization,
4. melt crystallization followed by drawing,
5. melt crystallization with tensile stress,
6. melt crystallization under high pressure, and
7. crystallization from concentrated solution followed by tensile elongation at an elevated temperature.

The density of most flexible chain polymers is greater in the crystalline state than in the liquid state and, therefore, an increase in crystallinity leads to an increase in density. The specific heat of fusion (Joules per gram) increases with the amount of crystallinity, but it also depends on the morphology [1]. The diffraction pattern obtained when an X-ray beam impinges on a polymer depends on the crystallinity. Large changes in intensity (with a small change in scattering angle at various mean scattering angles) are associated with the crystallites, while a broad, less intense band is characteristic of the amorphous portions. The infrared and the Raman spectra for a particular polymer will show numerous characteristic bands, each dependent on a particular normal vibration [2]. The frequency positions of many of these depend on the chain conformation; therefore, an increase in crystallinity will lead to an increase in intensities for bands (or portions of bands) associated with the conformation adopted by the chains in the crystallite. Solid state carbon-13 nuclear magnetic resonance gives a spectrum that depends on the bonding and the environment for the carbon atoms present in the structure. Due to magnetic environmental differences in crystalline and amorphous portions of a crystallized polymer, separate characteristic resonance peaks will be observed for some of the carbon atoms present [3]. The intensities of these will change with a change in the crystallinity. All of the physical methods mentioned here have been used to some degree to measure the crystallinity of polymer samples. Differences obtained using two different methods of measurement with the same sample arise due to:

1. The type of crystallinity (weight % or volume %) measured,
2. calibration errors, and
3. the dependence of a given method on additional factors, such as the orientation of the chains in the sample.

8.2 Mechanical Behavior

8.2.1 Tensile Stress-strain Curves

The tensile stress plotted versus the extension ratio, γ, in the sample is given in Fig. 8.1 for three different flexible homopolymers and one blend. Note the large differences in the scales of stress and extension ratio used in the three plots. The homopolymers are: A) atactic polystyrene, which is glassy [4], B) polyethylene, which is crystallizable with rubbery intercrystalline chain portions [5], and C) lightly crosslinked natural rubber [6]. The blend, HIPS, contains small rubber particles dispersed in polystyrene [4]. The curve for the glassy material, polystyrene, shows that the strain elicited by the stress at 25 °C remains low, yet

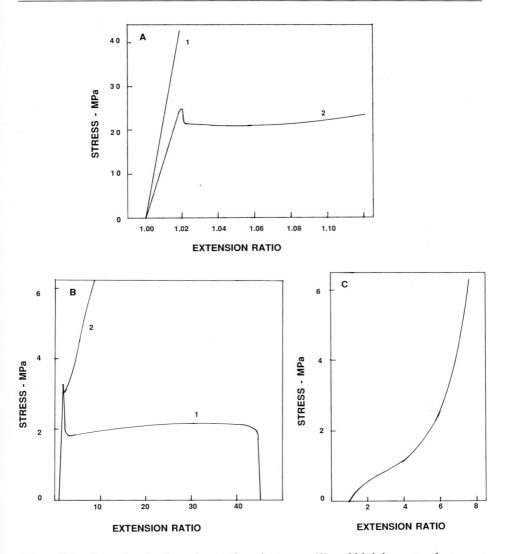

Figure 8.1   Stress/strain plots: A. atactic polystyrene (1) and high impact polystyrene (2) [From: McCrum, N.G., Buckley, C.P. and Bucknall, C.B. (1988) *Principles of Polymer Engineering*. Oxford: Oxford University Press, p. 183]; B. polyethylene drawn at 90 °C, as polymerized at 50 °C (1) and melt-recrystallized (2) [From: Smith, P., Chanzy, H.D. and Rotzinger, B.P. (1987) J. Mat. Sci. *22*, 523]; and C. lightly crosslinked natural rubber at 50 °C [From: Treloar, L.R.G. (1944) Trans. Faraday Soc. *40*, 49].

failure occurs at relatively large stress values. The crosslinked rubber responds initially to the stress with a large deformation, followed by stress hardening and, finally, fracture. The blend has a behavior that differs from that for both components separately, with yielding and some extension taking place. The response of polyethylene to a tensile stress is found to depend strongly on the morphology. The sample, as polymerized at 50 °C, shows yielding followed by about a fortyfold extension at 90 °C prior to fracture. As polymerized material recrystallized from the melt , it shows some drawing at 90 °C but fractures at about eightfold extension. The polystyrene and HIPS samples both sustain a higher stress before fracture, but reach only a fraction of the extension ratio as compared to that for either rubber or polyethylene.

The stress/strain curves for these four general types of polymer systems depend on the rate of strain and the measurement temperature. At temperatures below T_g, the stress/strain curve for a partially crystalline polymer resembles that for a glassy polymer; at temperatures above T_g, the yield point occurs at lower stress with yielding increasing with temperature. Resistance to crack initiation was found to increase with crystal thickness [7].

8.2.2   Tensile Strength

For flexible chain polymers, the tensile strength is highest for glassy liquids (such as polystyrene), intermediate for cold drawn crystallizable polymers (such as polyethylene) and lowest for crosslinked rubbers. Rigid crosslinked systems, such as cured epoxy resins, have higher tensile strengths than flexible chain glassy polymers; the tensile strength of a cured epoxy resin is increased about fivefold by fabric reinforcement and about ten- to twentyfold by filament winding [8].

The tensile strength of a crystallizable polymer, such as polyethylene, increases with the morphology in the order:

1. Filtered mats of single lamellas, hedrites and spherulites from solution, extended chain spherulites from the melt,
2. hedrites and spherulites quiescently crystallized from the melt,
3. melt crystallized polymer under extensional flow, and
4. ultradrawn crystallized gel.

Groups of single lamellas and individual hedrites and spherulites have only van der Waals forces holding together the individual structures making up the sample, leading to low tensile strength. Extended chain spherulites grown under pressure from the melt are brittle for the same reason. Melt-crystallized hedrites and spherulites, cold drawn upon application of stress and at fracture, contain a network of extended chains and crystallites with the chains oriented principally in the same direction. In ultradrawn materials, the chains are extended and oriented in the same direction, with chain ends appearing at random positions within the fiber. Ultradrawn polyethylene has a tensile strength about eight times higher than cold drawn material [9]. Poly(1,4-benzamide) fiber has a tensile strength that is about eight times larger than ultradrawn polyethylene.

As described in Chapter 7, porous polyethylene film can be prepared by biaxial orientation of material extruded from concentrated solution [10]. Nonporous samples result if annealing at 120 °C is carried out prior to biaxial stretching; annealing improves connectivity between the fibrils in the films without melting the crystals present. The porous film has a lower tensile strength in comparison to the nonporous film given the same amount of biaxial orientation.

As discussed in Chapter 6, shearing of thin films of liquid crystalline mesophases causes the formation of bands running perpendicular to the shearing direction. The size of the bands present, controlled by changing the rate of solvent evaporation, is found to affect the mechanical properties of 2-hydroxypropylcellulose films [11]. A fourfold increase in band thickness (from 2.95 to 12.50 μm) causes an increase in the stress at break along the shearing direction and a twofold decrease in this parameter transverse to the shearing direction. On deformation, a large amount of molecular reorientation results, causing smaller bands to appear when the band thickness is large.

Blending of a crystallizable polymer with another crystallizable polymer, or with a rubber, can change the tensile strength. Blending of polypropylene with polyethylene or polybutene-1 leads to an improvement in fracture stress when thin films of the two components are sandwiched together, subjected to a tensile stress during melt crystallization and annealed [12–14].

Blending with a rubber or with another crystallizable polymer can result in an increase in tensile strength for crystallized polymers with T_g near to or above the testing temperature. The tensile strength is a maximum for injection-molded isotactic poly(butene-1) (PB-1)/isotactic polypropylene at 25 wt % PB-1 [15]. A shift in this maximum to 50 wt % occurs after annealing for one hour at 145 °C. This tensile strength enhancement does not occur for compression-molded blends due to the occurrence of macrophase separation during molding.

The formation of miscible cured blends can result in improved tensile strength [16]; cured blends of (78/28%) ethylene/vinyl acetate copolymer with branched polyethylene show a maximum in the tensile strength for equal parts of the two constituents.

Enhancement of the mechanical properties in blends and composites is aided by good adhesion between the polymer matrix and the dispersed blend or composite particles. This behavior is observed for materials containing mica and for rubber/carbon black mixtures.

The tensile strength of an elastomeric triblock copolymer of styrene/(ethylene/butylene)/styrene is increased by blending with a fibrous liquid crystalline polymer [17]; although the two components in the blend are highly incompatible, T_g of the elastomer shifts with the composition.

8.2.3   Impact Strength

Although flexible chain glassy polymers will sustain a large statically applied stress prior to fracture, they are brittle and have low impact strengths as compared to rubbers and crystallized polymers. The presence of a rubbery component as inclusions in a glassy matrix in phase separated block copolymers and blends raises the impact strength while lowering the tensile strength [4].

The addition of methacrylated butadiene/styrene (MBS) to blends of polycarbonate (PC) with poly(methyl methacrylate) or with styrene/acrylonitrile copolymer (SAN), causes a significant increase in toughness [18]. The preparation and morphology of these blends is discussed in Chapter 6. A 60/30/10 blend of PC, SAN25 and MBS has an impact strength about seven times larger than that for a blend of PC with SAN25. When SAN17.5 or SAN34 are used in place of SAN25, increases in impact strength of about sixfold occur upon MBS addition. The impact strength depends on the method of mixing the three components, as is seen from the data given in Table 8.1. When the SAN25 is first mixed with MBS, followed by mixing with PC, the tensile strength is larger by about 30% than that of samples prepared using simultaneous mixing or either of the other two binary mixing sequences. This increase correlates with a larger amount of MBS residing in the brittle SAN25 phase.

In an investigation of the effect of mineral fillers on the impact strength of polypropylene, it was found that filler dispersion and nucleating ability were important variables [19]. Higher nucleation and poorly dispersed filler led to a lower impact strength.

Improvement of the impact strength depends on the dispersed phase particle size, the rubbery material being used and the adhesion between the particles and the matrix. The optimum size is usually around 0.1–2 μm in diameter [20]. The variables affecting the size and distribution of the dispersed particles include: elastomer content, compatibility, the processing method and the viscosities of the two components. The presence of subinclusions appear to enhance the impact strength.

Impact resistance and higher resistance to rupture from internal pressure is improved by the biaxial orientation that occurs in blow molding [4].

Table 8.1 Impact Strength and Distribution of MBS Particles in 60/30/10 Blends of PC, SAN25 and MBS[a]

Mixing sequence	MBS distribution, %			Impact Strength J m[-1]
	PC phase	SAN25 phase	interface	
Simultaneous	35	5	60	496
[PC/SAN25]+MBS	5	5	90	464
[PC/MBS]+SAN25	55	5	40	448
[SAN25/MBS]+PC	15	25	60	630

[a] Ref. [18].

Note: Three binary mixing sequences were used; these are denoted in the table by the pair mixed first being given in brackets followed by the third component added last.

8.2.4   Compressive Strength

Materials with extended chain structures, such as polybenzothiazole and polybenzoxazole, have very high tensile strengths but have compression strengths that are ten times lower [21]. Failure occurs by bond bending and breaking due to strain localization in the kink bands that form.

The interfacial shear strength of carbon fiber/amine cured epoxy composites under compression increases with pretreatment of the fibers [22]. Coating of the fibers with uncured epoxy prior to mixing the components gave a composite that fractured by compressive failure of the fibers; electrochemical oxidation of the fiber surface resulted in a composite that showed fiber microbuckling that progressed along the direction of maximum shear stress, while the use of untreated fibers resulted in delamination and delamination buckling.

8.2.5   Yielding

The amount of tensile elongation of crystallized polymer samples occurring prior to fracture at temperatures between T_g and T_m depends on the sample morphology and appears to be affected by the number of chain entanglements present [23]. Melt-crystallized spherulitic samples from flexible chain polymers can be drawn to ultimate elongations of about threefold to eightfold, depending on the temperature, the polymer and the molecular weight. Films crystallized from concentrated solution or obtained during polymerization can sustain much higher draw amounts with up to 130-fold being reported for high molecular weight polyethylene film [5, 24, 25]. Gel-spinning, followed by ultradrawing, is another method used to obtain a high degree of drawing (up to about seventyfold)

[23]. This process involves spinning/extrusion of a semidilute solution of the polymer; the as-spun quenched filaments, with or without solvent present, are then ultradrawn. Gel-spinning has been applied to polyethylene, polypropylene, polyacrylonitrile and poly(vinyl alcohol). Polyethylene fibers prepared from spherulitic samples can sustain some further elongation (on the order of 50%) prior to fracture, while ultradrawn fibers show little further elongation (2%) [9]. Poly(1,4-benzamide), a semirigid chain polymer, in fiber form exhibits a small elongation prior to break (6%).

The tensile yield stress of ethylene/butene-1 copolymers crystallized from the melt and from solution was found to increase with an increase in lamella thickness but to change irregularly with a change in the crystalline fraction in the range of 0.33 to 0.74 [26].

The presence of large amounts of a filler (40%), such as mica, in isotactic polypropylene increases the ductility of the material at 25 °C [27]. A possible explanation given is that a developing crack front is held together (pinned) by the mica particles, allowing a longer time for plastic deformation of the matrix near the front.

The effects of band size on the tensile strength of 2-hydroxypropyl-cellulose was discussed above. Changes in the strain at break, particularly transverse to the shearing direction, also occur. The strain at break in the transverse direction increases eighteenfold for a fourfold increase in band thickness. In the parallel direction, the strain at break decreases for thicker bands.

The ultimate elongation of a glassy polymer can be increased by blending or by block copolymer formation with a rubbery polymer with the latter in the lesser amount and forming the inclusions [8]. An example of this is the increase in ultimate elongation for HIPS as compared to that for atactic polystyrene (see Fig. 8.1A).

8.3   Optical Properties

Glassy polymers without additives or voids are clear and transmit visible light; many of these absorb in the ultraviolet region. The transmission of visible light through thick films of crystallized polymers varies with the morphology and in some cases with the sample orientation. Spherulite-containing samples can scatter an appreciable amount of light. Both quenched films and highly oriented films of crystallizable polymers, which do not absorb in the visible, should show a greater light transmittance than semicrystalline ones. Similarly, ultradrawn polyethylene films, with the surface coated with ethylene glycol in order to reduce surface scattering, have a transmittance for visible light (wavelengths of 400–800 nm) of greater than 90% [28].

The transmission of polarized light by crystallized polymers and liquid crystalline mesophases was discussed in Chapters 3 and 4. Extinction occurs when the optic axis of the polymer chain or mesogenic group is parallel to either the polarizer or analyser directions. Isotropic polymer fluids, rubbers and glasses, appear dark through crossed polaroids. Orientation of the chains in these, caused by the fabrication technique or by stress application, can result in the appearance of birefringence effects [29]. The unoriented portions of the sample will remain dark.

The absorption of visible light by polydiacetylene single crystals depends on the wavelength, the incident polarization orientation and the temperature [30, 31]. Single crystals of the polydiacetylene with:

$$-(CH_2)_4-O-(C=O)-NH-\text{⟨⟩}$$

side chains absorb in the visible region with a strong absorption at 610 nm wavelength when the light polarization is parallel to the chain axis; they have a low absorption at all wavelengths in the visible region when the polarization is perpendicular [30]. Single crystals of polydiacetylene with:

$$-(CH_2)_4-O-(C=O)-NH-C_2H_5$$

side chains absorb in the blue region below 115 °C and in the red region above that temperature when the polarization is parallel to the chain direction [31].

Infrared absorption depends on the wavelength, the polymer and the chain orientation in regard to the polarization direction of the light. Infrared irradiation interacts with a dipole which has a component oscillating parallel to the polarization direction of the wave. If the dipole oscillates in a direction perpendicular to the polarization direction, no interaction between the wave and the oscillating dipole will occur. Oriented samples will have different infrared intensities than randomly ordered ones [2].

Nonlinear optical (NLO) behavior is observed for polymers with oriented chains, such as found in single crystal films [9]. Subjecting a thin, multicrystallite film of a dipole-containing polymer to a high voltage (10^6 V/cm) above T_g, known as poling, followed by quenching while the high voltage is still applied, introduces small amounts of chain orientation and NLO response [32, 33].

8.4   Electrical Properties

The electrical resistivity of a polymer depends on the polarity of the various groups of atoms in the chain. Polyethylene is nonpolar and has a relatively high resistivity, while polymers containing functional groups in the main chain or on the side chains are usually polar. The log resistivity of miscible blends of low density polyethylene with a 72%/28% ethylene/(vinyl acetate) copolymer shows a large decrease when the amount of copolymer present is 20–40%, with little additional change above 40% [16].

The electrical polarization increases with the degree of orientation in a polymer sample. Therefore, a crystallized polymer will show a higher polarization than the same polymer in a completely amorphous condition; fibers will have a higher polarization than spherulitic samples [34].

The conductivity of polyacetylene is affected by the morphology [8, 35]. A lowering of the chain planarity, crosslinking, a low crystallinity, or the presence of voids lead to a lower conductivity. Single crystals are expected to have the highest conductivity. Changes in the chemical structure are also very important. Electrical breakdown of a polar polymer can be caused by defects, such as cracks, voids and foreign inclusions [36].

References

1. Wunderlich, B. (1973) *Macromolecular Physics, Vol. 1*. New York: Academic Press, pp. 380-408.
2. Painter, P.C., Coleman, M.M. and Koenig, J.L. (1982) *The Theory of Vibrational Spectroscopy and its Application to Polymeric Materials*. New York: Wiley.
3. Schilling, F., Bovey, F.A., Tonelli, A.E., Tseng, S. and Woodward, A.E. (1984) Macromols. *17*, 728.
4. McCrum, G., Buckley, and Bucknall, C. (1988) *Principles of Polymer Engineering*. Oxford: Oxford University Press, pp. 183, 311–313.
5. Smith, P., Chanzy, H. and Rotzinger, B.P. (1987) J. Mat. Sci. *22*, 523.
6. Treloar, L.R.G. (1944) Trans. Faraday Soc. *40*, 49.
7. Barry, D.B. and Delatycki, O. (1992) Polymer *33*, 1261.
8. Allcock, H. and Lampe, F. (1990) *Contemporary Polymer Chemistry*, 2nd Ed. Englewood Cliffs, NJ: Prentice-Hall.
9. Sperling, L.H. (1992) *Introduction to Physical Polymer Science*, 2nd Ed. New York: Wiley-Interscience.
10. Gerrits, N.S.J.A. and Lemstra, P.J. (1991) Polymer *32*, 1770.
11. Wang, J. and Labes, M.M. (1992) Macromols. *25*, 5790.
12. Gross, B. and Petermann, J. (1984) J. Mat. Sci. *29*, 105.
13. Jaballah, A., Rieck, U., and Petermann, J. (1990) J. Mat. Sci. *25*, 3105.
14. Petermann, J., Xu, Y., Loos, J. and Yang, D. (1992) Makromol. Chem. *193*, 611.

15. Lee, M.-S. and Chen, S.-A. (1993) Polym. Eng. Sci. *33*, 686.
16. Ray, I. and Khastgir, D. (1993) Polymer *34*, 2030.
17. Verhoogt, H., Langehan, H.C., Van Dam, J. and Posthuma DeBoer, A. (1993) Polym. Eng. Sci. *33*, 754.
18. Cheng, T.W., Keshkula, H. and Paul, D.R. (1992) Polymer *33*, 1606.
19. McGenity, P.M., Hooper, J.J., Paynter, C.D., Riley, A.M., Nutbeen, C., Elton, N.J. and Adams, J.M. (1992) Polymer *33*, 5215.
20. Sawyer, LC. and Grubb, D.T. (1987) *Polymer Microscopy*. London: Chapman and Hall.
21. Martin, D.C. and Thomas, E.L. (1991) J. Mat. Sci. *26*, 5171.
22. Drzal, L.T. and Madhukar, M. (1993) J. Mat. Sci. *28*, 569.
23. Lemstra, P.J., van Aerle, N.A.J.M. and Bastiaansen, C.W. (1987) Polym. J. *19*, 85.
24. Smith, P., Boudet, A. and Chanzy, H. (1985) J. Mat. Sci. Letts. *4*, 13.
25. Smith, P. Lemstra, P.J., Pijpers, J.P.L. and Kiel, A.M. (1981) Colloid and Polymer Sci. *259*, 1070.
26. Darras, O. and Segula, R. (1993) J. Polym. Sci.: Polym. Phys Ed. *31*, 759.
27. Xavier, S.X., Schultz, J.M. and Friedrich, K. (1990) J. Mat. Sci. *25*, 2411.
28. Bastiaansen, C., Schmidt, H.-W., Nishino, T. and Smith, P. (1993) Polymer *34*, 3951.
29. Campbell, D. and White, J.R. (1989) *Polymer Characterization*. London: Chapman and Hall, pp. 271–299.
30. Thakur, M. and Meyler, S. (1985) Macromols. *18*, 2341.
31. Tanaka, H., Thakur, M., Gomez, M.A., and Tonelli, A.E. (1987) Macromols. *20*, 3094.
32. Sohn, J.E., Singer, K.D., Kuzyk, M.G., Holland, W.R., Katz, H.E., Dirk, C.W., Schilling, M.L. and Comizzoli, R.B. (1988) NATO ASI Series E *162*, 291.
33. Ulrich, D.R. (1988) NATO ASI series E *162*, 299.
34. Daniels, C.A. (1989) *Polymers:Structure and Properties*. Lancaster, PA: Technomic.
35. Frommer, J.E. and Chance, R.R. (1986) in *Encyclopedia of Polymer Science and Engineering*, 2nd Ed., Vol. 5. New York: Wiley, p. 462.
36. Mathes, K.N. (1986) in *Encyclopedia of Polymer Science and Engineering*, 2nd Ed., Vol. 5. New York: Wiley, p 507.

Index